EAI/Springer Innovations in Communication and Computing

Series Editor

Imrich Chlamtac
European Alliance for Innovation
Ghent, Belgium

Editor's Note

The impact of information technologies is creating a new world yet not fully understood. The extent and speed of economic, life style and social changes already perceived in everyday life is hard to estimate without understanding the technological driving forces behind it. This series presents contributed volumes featuring the latest research and development in the various information engineering technologies that play a key role in this process.

The range of topics, focusing primarily on communications and computing engineering include, but are not limited to, wireless networks; mobile communication; design and learning; gaming; interaction; e-health and pervasive healthcare; energy management; smart grids; internet of things; cognitive radio networks; computation; cloud computing; ubiquitous connectivity, and in mode general smart living, smart cities, Internet of Things and more. The series publishes a combination of expanded papers selected from hosted and sponsored European Alliance for Innovation (EAI) conferences that present cutting edge, global research as well as provide new perspectives on traditional related engineering fields. This content, complemented with open calls for contribution of book titles and individual chapters, together maintain Springer's and EAI's high standards of academic excellence. The audience for the books consists of researchers, industry professionals, advanced level students as well as practitioners in related fields of activity include information and communication specialists, security experts, economists, urban planners, doctors, and in general representatives in all those walks of life affected ad contributing to the information revolution.

Indexing: This series is indexed in Scopus, Ei Compendex, and zbMATH.

About EAI

EAI is a grassroots member organization initiated through cooperation between businesses, public, private and government organizations to address the global challenges of Europe's future competitiveness and link the European Research community with its counterparts around the globe. EAI reaches out to hundreds of thousands of individual subscribers on all continents and collaborates with an institutional member base including Fortune 500 companies, government organizations, and educational institutions, provide a free research and innovation platform.

Through its open free membership model EAI promotes a new research and innovation culture based on collaboration, connectivity and recognition of excellence by community.

More information about this series at http://www.springer.com/series/15427

EAI/Springer Innovations in Communication and Computing

Series Editor

Imrich Chlamtac
European Alliance for Innovation
Ghent, Belgium

Editor's Note

The impact of information technologies is creating a new world yet not fully understood. The extent and speed of economic, life style and social changes already perceived in everyday life is hard to estimate without understanding the technological driving forces behind it. This series presents contributed volumes featuring the latest research and development in the various information engineering technologies that play a key role in this process.

The range of topics, focusing primarily on communications and computing engineering include, but are not limited to, wireless networks; mobile communication; design and learning; gaming; interaction; e-health and pervasive healthcare; energy management; smart grids; internet of things; cognitive radio networks; computation; cloud computing; ubiquitous connectivity, and in mode general smart living, smart cities, Internet of Things and more. The series publishes a combination of expanded papers selected from hosted and sponsored European Alliance for Innovation (EAI) conferences that present cutting edge, global research as well as provide new perspectives on traditional related engineering fields. This content, complemented with open calls for contribution of book titles and individual chapters, together maintain Springer's and EAI's high standards of academic excellence. The audience for the books consists of researchers, industry professionals, advanced level students as well as practitioners in related fields of activity include information and communication specialists, security experts, economists, urban planners, doctors, and in general representatives in all those walks of life affected ad contributing to the information revolution.

Indexing: This series is indexed in Scopus, Ei Compendex, and zbMATH.

About EAI

EAI is a grassroots member organization initiated through cooperation between businesses, public, private and government organizations to address the global challenges of Europe's future competitiveness and link the European Research community with its counterparts around the globe. EAI reaches out to hundreds of thousands of individual subscribers on all continents and collaborates with an institutional member base including Fortune 500 companies, government organizations, and educational institutions, provide a free research and innovation platform.

Through its open free membership model EAI promotes a new research and innovation culture based on collaboration, connectivity and recognition of excellence by community.

More information about this series at http://www.springer.com/series/15427

Shenglin Mu • Li Yujie • Huimin Lu
Editors

4th EAI International Conference on Robotic Sensor Networks

Springer

EAI
RESEARCH MEETS INNOVATION

Editors
Shenglin Mu
Ehime University
Matsuyama, Japan

Li Yujie
Fukuoka University
Fukuoka, Japan

Huimin Lu
Kyushu Institute of Technology
Kitakyushu, Japan

ISSN 2522-8595 ISSN 2522-8609 (electronic)
EAI/Springer Innovations in Communication and Computing
ISBN 978-3-030-70450-6 ISBN 978-3-030-70451-3 (eBook)
https://doi.org/10.1007/978-3-030-70451-3

This Springer imprint is published by the registered company Springer Nature Switzerland AG
The registered company address is: Gewerbestrasse 11, 6330 Cham, Switzerland

Shenglin Mu • Li Yujie • Huimin Lu
Editors

4th EAI International Conference on Robotic Sensor Networks

Springer

EAI
RESEARCH MEETS INNOVATION

Editors
Shenglin Mu
Ehime University
Matsuyama, Japan

Li Yujie
Fukuoka University
Fukuoka, Japan

Huimin Lu
Kyushu Institute of Technology
Kitakyushu, Japan

ISSN 2522-8595 ISSN 2522-8609 (electronic)
EAI/Springer Innovations in Communication and Computing
ISBN 978-3-030-70450-6 ISBN 978-3-030-70451-3 (eBook)
https://doi.org/10.1007/978-3-030-70451-3

This Springer imprint is published by the registered company Springer Nature Switzerland AG
The registered company address is: Gewerbestrasse 11, 6330 Cham, Switzerland

Preface

We are delighted to introduce the proceedings of the 2020 European Alliance for Innovation (EAI) International Conference on Robotic Sensor Networks ROSENET2020. The theme of ROSENET 2020 was "Cognitive Robotics for Smart Society." These proceedings highlight selected papers presented at the 4th EAI International Conference on Robotic Sensor Networks, which was held virtually. Today, the integration of artificial intelligence and the Internet of Things has become a topic of growing interest for both researchers and developers from academic fields and industries worldwide, and artificial intelligence is poised to become the main approach pursued in next-generation IoTs research.

The rapidly growing number of artificial intelligence algorithms and big data devices has significantly extended the number of potential applications for IoT technologies. However, it also poses new challenges for the artificial intelligence community. The aim of this conference is to provide a platform for young researchers to share the latest scientific achievements in this field, which are discussed in these proceedings.

The technical program of ROSENET 2020 consisted of 11 full papers from 34 submissions. Apart from the high-quality technical paper presentations, the technical program also featured three keynote speeches by Prof. Mengchu Zhou, New Jersey Institute of Technology, USA, and Prof. Xizhao Wang, Shenzhen University, China.

Coordination with the steering chair, Imrich Chlamtac, was essential for the success of the conference. We sincerely appreciate his constant support and guidance. It was also a great pleasure to work with such an excellent organizing committee team. We appreciate their hard work in organizing and supporting the conference, in particular the technical program committee, led by our program chair, Dr. Shenglin Mu, Dr. Jože Guna, and Dr. Shota Nakashima, who completed the peer-review process of technical papers and created a high-quality technical program. We are also grateful to conference managers, Lukas Skolek for his support, and all the authors who submitted their papers to the ROSENET 2020 conference and special sessions.

We strongly believe that ROSENET conferences provide a good forum for all researcher, developers, and practitioners to discuss all science and technology aspects that are relevant to robotics and cognitive Internet of Things. We also expect that the future ROSENET conferences will be as successful and stimulating, as indicated by the contributions presented in this volume.

Matsuyama, Japan — Shenglin Mu
Fukuoka, Japan — Li Yujie
Kitakyushu, Japan — Huimin Lu

Conference Organization

Steering Committee

Imrich Chlamtac, Bruno Kessler Professor, University of Trento, Italy
Huimin Lu, Kyushu Institute of Technology, Japan

Organizing Committee

General Chair
Huimin Lu, Kyushu Institute of Technology, Japan
General Co-Chairs
Hao Gao, Nanjing University of Posts and Telecommunications, China
TPC Chair and Co-Chair
Li Yujie, Fukuoka University, Japan
Sponsorship and Exhibit Chair
Joze Guna, University of Ljubljana, Slovenia
Local Chair
Yin Zhang, Zhongnan University of Economics and Law, China
Workshops Chair
Guangxu Li, Tianjin Polytechnic University, China
Publicity & Social Media Chair
Junwu Zhu, Yangzhou University, China
Publications Chair
Shota Nakashima, Yamaguchi University, Japan
Web Chair
Shenglin Mu, Ehime University, Japan
Posters and PhD Track Chair
Xin Jin, Beijing Electronic Science and Technology Institute, China
Panels Chair

Zhenlong Du, Nanjing Tech University, China
Demos Chair
Dong Wang, Dalian University of Technology, China
Tutorials Chairs
Xing Xu, University of Electrical Science and Technology of China, China

Technical Program Committee

Shiming Ge, Chinese Academy of Sciences, China
Haitao Cheng, Nanjing University of Posts and Telecommunications, China
Csaba Beleznai, AIT Austrian Institute of Technology, Austria
Mario G. C. A. Cimino, University of Pisa, Italy
Lu Tao, Wuhan Institute of Technology, China
Wankou Yang, Southeast University, China
Mohammed A. B. Mahmoud, Beijing Institute of Technology, China
Tongwei Ren, Nanjing University, China
Zhibin Yu, Ocean University of China, China
Ting Wang, Nanjing Tech University, China
Quan Zhou, Nanjing University of Posts and Telecommunications, China
Haiyong Zheng, Ocean University of China, China
Yaochen Li, Xi'an Jiaotong University, China
Xuebin Qin, Xi'an university of Science and technology, China
Fei Xu, Xi'an Technological University, China
Chiew Foong Kong, University of Nottingham Ningbo China, China
Xin Kang, Tokushima University, Japan
Jianpin Gou, Jiangsu Unversity, China
Hua Chen, Jiangxi Normal University, China
Xipeng Pan, Beijing University of Posts and Telecommunications, China
Fengquan Zhang, North China University of Technology, China
Song Deng, Nanjing University of Posts and Telecommunications, China
Mei Wang, Xi' an University of Science and Technology, China
Yi Chen, Nanjing Normal University, China
Yichun Wang, University of Sheffield, UK
Zongyuan Ge, Monash University, Australia
Ning Tan, Sun Yat-sen University, China
Yongjun Zhang, Guizhou University, China
Huiju Wang, Zhongnan University of Economics and Law, China
Jialiang Shen, iShare, Australia
Liu Xu, Yangzhou University, China
Keliang Jia, Shandong University of Finance and Economics, China
Konrad Karcz, Ludwig Maximilian University Munich, Germany
Qiufeng Li, Nanchang Hangkong University, China
Jianwu Wan, Nanyang Technological University, Singapore

Contents

A Mobile Robotic System for Rescue and Surveillance in Indoor Environment

Phan Van Vinh, Tran Minh Thanh, Vo Duy Khang, Ngo Hoang An, and Tran Duy Nhat

Abstract Recently, research activities related to Internet of Things (IoT) have been focused by many researchers in all aspects of human lives that made the rapid changes to the traditional model of a rescue and surveillance system. In this context, the development of an intelligent rescue and surveillance system with multi-functional robots is generally adopted to mitigate the accidents in rescue activities in dangerous situations. Therefore, in this paper, we design and implement a mobile robotic system that can handle rescue and surveillance tasks efficiently. With a mobile robot deployed in the given area to collect the real-time data of the target environment and get live video feedback, the system can not only measure the environmental conditions that may affect the health of the rescue team but also detect a victim being in the affected area. It saves precious time in rescue activities and helps the rescue team to plan their work more efficiently. The proposed system can be controlled and managed remotely via a friendly Web-based user interface.

Keywords IoT · Mobile robot · Rescue · Surveillance · Object detection

1 Introduction

Nowadays, Internet has become one of the most important aspects of our lives. Everyone uses Internet for almost all of their life activities. Internet of Things (IoT) is a network of devices such as cameras, vehicles, home appliances, and other

P. Van Vinh (✉) · T. M. Thanh · V. D. Khang · N. H. An
School of Computing and Information Technology, Eastern International University, Binh Duong City, Vietnam
e-mail: vinh.phan@eiu.edu.vn; thanh.tran.k2set1@eiu.edu.vn; khang.vo.k3set@eiu.edu.vn; an.ngo.set14@eiu.edu.vn

T. D. Nhat
School of Engineering, Eastern International University, Binh Duong City, Vietnam
e-mail: nhat.tran@eiu.edu.vn

S. Mu et al. (eds.), *4th EAI International Conference on Robotic Sensor Networks*, EAI/Springer Innovations in Communication and Computing,
https://doi.org/10.1007/978-3-030-70451-3_1

electronic devices which can connect, exchange, and transfer data together. One of the most widely used IoT applications is surveillance and rescue system.

A surveillance system with CCTV (closed circuit television) cameras or other IP cameras is used to monitor sensitive areas, such as outdoor areas (public places, garden, front of a house, etc.) and indoor areas (warehouse, home, garage, etc.). However, such cameras have some drawbacks and constraints such as they usually have blind spots, and also to cover an entire area, many cameras are required to reach every nook and corner. A rescue system helps save the life of victims who get stuck in dangerous situations. In such case, a rescue team cannot cover the entire affected area, especially the areas that are difficult to access. Moreover, it is dangerous and risky to enter such areas without getting the prior information about the number of victims and their condition and the environmental conditions of the target area. Sometimes, due to lack of information, a rescuer may become a victim. To overcome this problem, deploying a mobile robot into the given area is one of the best solutions. The mobile robot mounted with video camera and sensors can go to the affected area and send all real-time information from the target environment to the rescue team. By analyzing the collected data, the rescue team can plan their work more efficiently.

However, building such kind of mobile robot with all necessary features is very expensive, complex, and time-consuming. One mobile robotic system for indoor surveillance was proposed in [1]. This robot can handle basic problems related to environment mapping, localization, and autonomous navigation. However, it cannot detect the presence of people in the given area. The authors in [2] proposed a surveillance robot that is able to avoid obstacles based on back propagation neural network. To reduce the cost in building robots, [3] proposed a mobile robot based on Android smartphone. A similar approach which uses Android smartphone with Arduino microcontroller was also proposed [4]. However, due to the hardware and software limitations, the above-mentioned approaches do not perform well in image processing tasks, such as human detection. To handle this problem, a deep learning algorithm with high performance of accuracy and speed can be considered, such as Region Convolutional Neural Network (RCNN) [5], Fast RCNN [6], Multi-scale Deep Convolutional Neural Network (MS CNN) [7], Faster RCNN [8], and YOLO [9]. MobileNet-V2 [10] is the state-of-the-art object detection algorithm with the best performance in accuracy. However, it needs to be run on computers with powerful graphic cards compared to other algorithms.

In this paper, we aim to design and implement a mobile robotic system with all necessary features at low cost. The mobile robot system can provide live video stream of the target area, detect human presence, and detect other environmental related information such as humidity, temperature, CO, and poisonous gas. The proposed system can be controlled and managed remotely via Web-based application. The contribution of this paper is as follows: (1) to design and implement a mobile robot prototype and test it in practical scenarios and (2) to implement a web-based application to control and manage the mobile robot system.

In what follows, we present the principal design of the proposed approach in Sect. 2 and implementation of results in Sect. 3. Finally, we make concluding remarks in Sect. 4.

2 Mobile Robotic System Design

In this section, we discuss the main design of the proposed system, including the system architecture, hardware and software design, and the prototype system model to help our readers understand clearly how our system works.

2.1 System Architecture

In this paper, we aim at designing a mobile robotic system to help rescue team in their works in a timely and efficient manner. Therefore, the system architecture should be simple, that is, easy to control and manage as shown in Fig. 1.

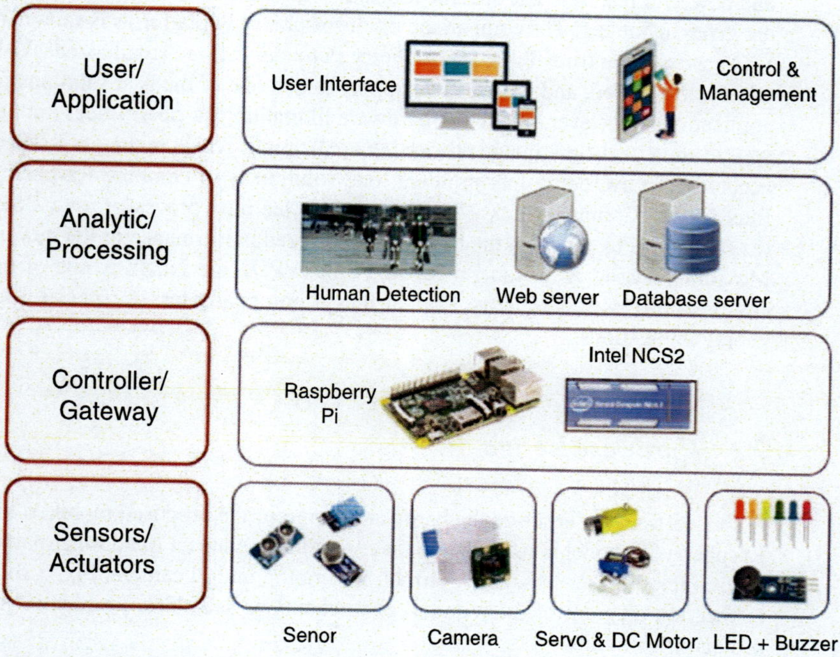

Fig. 1 The architecture of the proposed robotic system

Each mobile robot is embedded with multifunctional sensors to collect the ambient environmental data (e.g., temperature, humidity, gas, smoke, and distance to object) and a camera to get live image from the target area. All these data are sent to the center server for further processing. The ambient data are analyzed by using a fuzzy logic control algorithm to judge whether the target environment is safe or not, while the real-time images or videos are analyzed by using an objection detection algorithm to judge whether there is a human inside the target area. This information is very helpful to the rescue team in their work and to control the robot movement. The system can be accessed via web-based user interface to control the robot and observe the data environment. When the target environment becomes too dangerous or there is a victim in the given area, an alert or notification message will be displayed to notify the admin team about dangerous situations.

In our system, each mobile robot mounted with high-quality video camera needs to perform an algorithm to detect the defined objects in the target environment. Therefore, the main controller of each robot should be capable of processing real-time data at a very high speed and connect with multiple sensors and actuators.

At this time, there are several kinds of IoT-enabled devices with different capabilities and performance, such as Raspberry Pi models, Nviadia Jetson family, and Google Edge TPU Dev Board. Because of project requirements, we chose to use the Raspberry Pi model, a Linux-based high-performance computer with low cost and powerful features. The specifications of some well-known Raspberry Pi models are given in Table 1. The comparison of all Raspberry Pi platforms is shown in [11].

As we can see from Table 1, Raspberry Pi model that are small-sized, Wi-Fi and Bluetooth enabled, and with GPIO support can be one of the best solutions for IoT applications. However, due to the hardware limitation, Raspberry does not perform well in some real-time image processing applications which require high speed and accuracy. To overcome this problem, one solution is to combine Raspberry with Intel Neural Compute Stick 2 (NCS2), a deep learning processor on a USB stick (as shown in Fig. 2). With the hardware-optimized performance of the newest Intel Movidius Myriad X Vision Processing Unit (VPU), the NCS2 is one of the best combinations with Raspberry Pi 4 in image processing tasks, especially in deep neural applications.

2.2 Hardware Design

In this section, we describe the hardware design of the proposed mobile robot. The Raspberry Pi model is used as a main controller to connect to all other parts of the robot, including sensors, LED, buzzer, DC motor, and Pi camera. Fig. 3 shows the connection diagram of the required components on breadboard which is great for prototyping circuits.

Table 2 presents all the required hardware components of our mobile robot design.

Table 1 The specifications of some selected Raspberry Pi models

Raspberry Pi platform	Raspberry Pi Zero Wireless	Raspberry Pi 3 B	Raspberry Pi 3 B+	Raspberry Pi 4B
Processor	1 GHz single-core ARM11	1.2 GHz, Quad-Core Cortex A53	1.4 GHz, Quad-Core Cortex A53	1.5 GHz, Quad-Core Cortex A72
RAM	512 MB	1 GB	1 GB	1/2/4 GB
USB	1 × micro USB	4× USB2.0	4× USB2.0	2× USB3.0 + 2× USB2.0
Ethernet	–	10/100 Mbps	10/100 Mbps	1Gbps
Wi-Fi	802.11n	802.11n	2.4GHz and 5GHz 802.11 b/g/n/ac	2.4GHz and 5GHz 802.11 b/g/n/ac
Bluetooth	4.1	4.1	4.2	5.0
HDMI	Mini-HDMI	Yes	Yes	2 × micro HDMI
GPIO	40 pins	40 pins	40 pins	40 pins

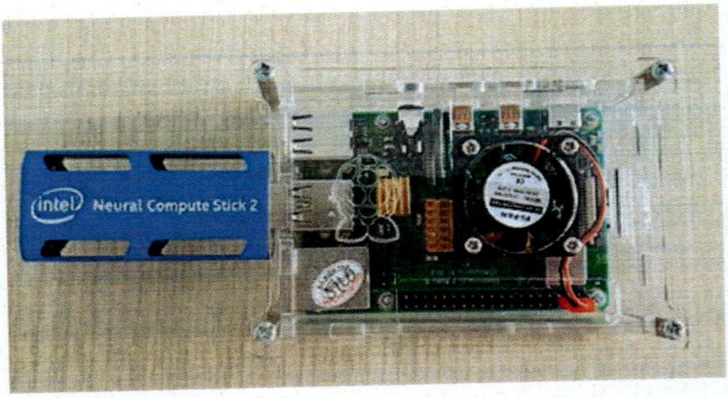

Fig. 2 Raspberry Pi 4 with Intel Neural Compute Stick 2

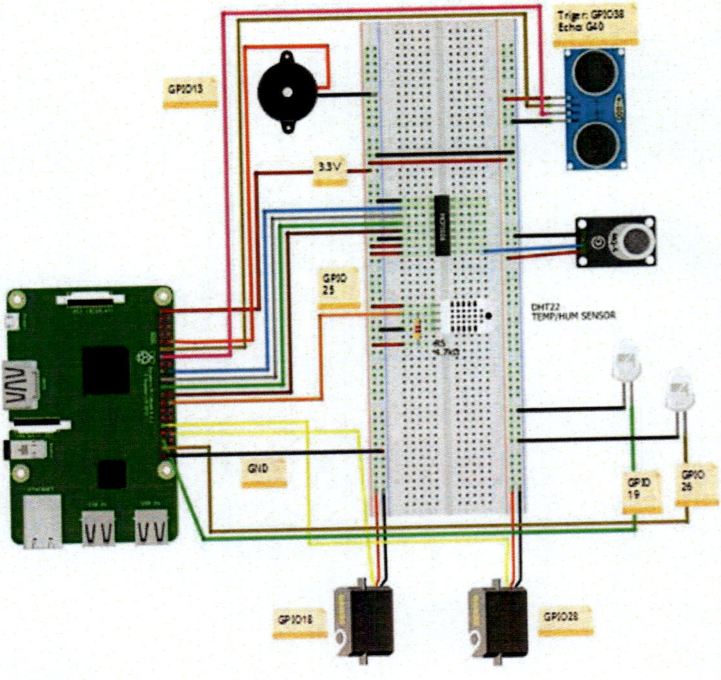

Fig. 3 The connection diagram of required parts for testing

When completing the prototyping step with breadboard, we have to make the robot work more permanent by designing the printed circuit board (PCB) of the mobile robot. Figure 4 illustrates the schematic diagram of all required component as follows:

- *Power supply block*: this includes the voltage reducer circuit with pulse power supply which can have maximum output of 5 V-3 A from the input voltage of 5–24 V DC to supply power for other components (e.g., Raspberry PI, sensors, and actuator).

Table 2 All hardware components of the designed mobile robot

Hardware components	Description
Raspberry Pi (3B+, 4B)	Main controller of the robot
Intel Neural Compute Stick 2	Deep learning device to accelerate edge devices like Raspberry Pi
MQ2 Gas/Smoke Sensors	Detecting LPG, smoke, alcohol, CH_4, CO,... in the air
DHT22	Measuring temperature and humidity
Ultrasonic Sensor HC-SR04	Measuring the distance accurately
Buzzer Module	Making a pip sound when needed
HD Pi Camera	Taking video or photo
SG90 Micro Servo Motor (×2)	Pan/tilt camera control
L298N Dual Motor Controller	Driving two DC motors
Robot tank car chassis	Robot tank chassis platform metal stainless steel with 2 DC motor

- *Motor driver block*: this includes Dual H-bridge driver chip L298N which can drive two DC motor and control the speed and direction of each motor independently through PWM (pulse width modulation). Input voltage is 5–30 V DC and maximum output is 2A for each bridge.
- *Actuator block*: this includes two servo motor, buzzer, and LED. The speed of servo motor can be controlled through PWM.
- *Sensor block*: this includes DHT11, SRF04, and MQ2 sensor for collecting environmental data.

The next step is to transfer the schematic diagram into a drawing of PCB (Fig. 5). The schematic will serve as a blueprint for laying out the traces and placing the components on the PCB. In addition, the PCB editing software can import all of the components, footprints, and wires into the PCB file, which will make the design process easier.

In this case, we use double-layer PCB, where the entire bottom layer is covered with a copper plane connected to ground. The thickness of power wires is 35 mil while that of data wires is 10 mil (0.254 mm). The positive traces are routed on top and connections to ground are made with through-holes or vias. Ground layers are good for circuits that are prone to interference because the large area of copper acts as a shield against electromagnetic fields. They also help dissipate the heat generated by the components (Fig. 6).

2.3 Software Design

Figure 7 shows the design of the web interface which is used to control the mobile robot movement and visualize the collected data. Users must have an account to log in into the system. There are two types of users: normal users who can only see

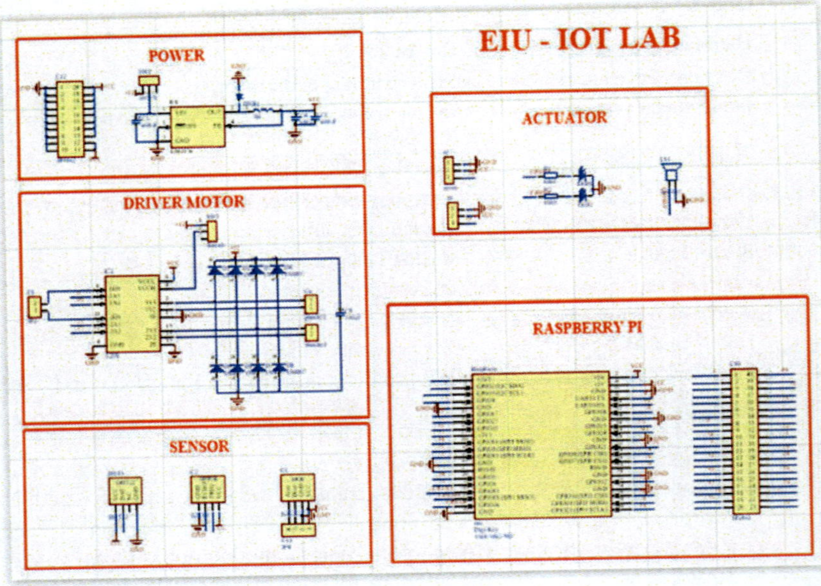

Fig. 4 The schematic view of electrical circuit diagram

the monitoring data and live-stream image and admin users who have full privileges to manage the whole system. The functionalities of some main pages in the web interface are as follows:

- *Robot Control* page: this controls the movement of robot and pan/tilt camera, observes the surrounding environment data and live stream from camera, records video or take a photo and save in gallery, enables/disables sensors or object detection feature, etc.
- *Data* page: this visualizes real-time environmental data acquisition in graph view and tabular view and displays the recorded videos and photos.
- *Location* page: this shows the location of the robot on the map via Google Map.

The collected data are stored in a MySQL database which includes the following table:

- User: stores the user information such as Id, name, role, e-mail, and contact number.
- Monitor_data: stores the collected data, including temperature, humidity, and MQ2.
- Obstacle_distance: stores distance value to obstacle.
- Photo: contains Id and name of a photo.
- Video: contains Id and name of a recorded video.

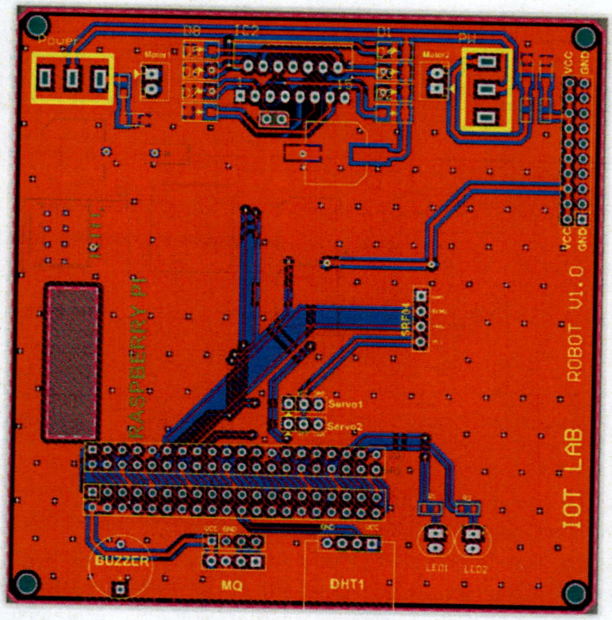

Fig. 5 The PCB layout of the designed mobile robot (2D view)

Fig. 6 The PCB layout of the designed mobile robot (3D view)

2.4 *Real-Time Human Detection*

Image classification and object detection are the major applications of machine learning. Image classification predicts the type or class of an object in an image, whereas object detection locates the presence of defined objects with a bounding box and assigns them a class label. Therefore, human detection is a specific application

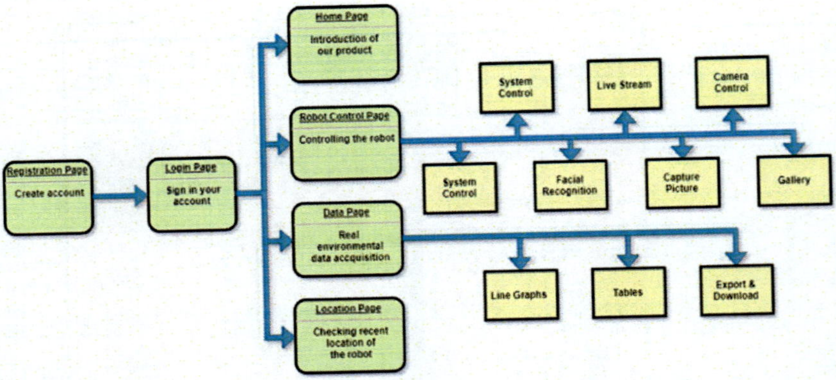

Fig. 7 The structure of the designed Web interface

of object detection to detect people in an image or video stream. In this paper, the mobile robot moves around the target area to find if there are people within that area. This information is very helpful for the rescue team to save the victim in a timely manner. At this time, for object detection, there are many state-of-the-art deep-learning algorithms, such as YOLOv3, MobileNet-V2, ResNet, and ResNet-50, which can be used with many popular machine-learning frameworks like TensorFlow, PyTorch, Caffe/Caffe2, Keras, MXNet, and others. MobileNet-V2 is the state-of-the-art object detection algorithm with the best performance in accuracy. However, it needs to be run on powerful graphic card (GPU) compared to other algorithms. SSD is designed to be independent of the base network, and so it can run on top of pretty much anything, including MobileNet. Even better, MobileNet+SSD uses a variant called SSDLite that uses depth-wise separable layers instead of regular convolutions for the object detection portion of the network. With SSDLite on top of MobileNet, you can easily get truly real-time results. In Sect. 3, we discuss about the experiments we conducted to verify the benchmarks of some selected edge devices which are attached with the Intel NCS2 USB stick and run MobileNet-V2 SSD and TensorFlow framework for object detection.

The process of human detection is shown in Fig. 8. We apply a human detection algorithm with deep-learning technique. The pre-trained Caffe object detection model is used for human detector. If there is a victim in the target area, a video or photo will be recorded and a logging entry is written to the database for further processing. The administration team will also be notified about the situation.

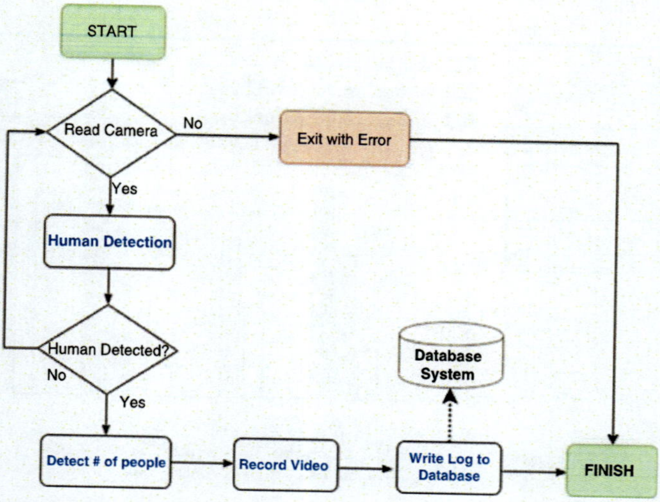

Fig. 8 The flowchart of human detection process

3 Experimental Results

3.1 Inference Benchmarks

To verify how the Intel NCS2 module can improve the performance of the edge devices in image processing tasks, we conduct some experiments with the inference benchmark testing. The experiments use SSD Mobilenet-V2 network model and TensorFlow framework, frame size of 400×400 with object detection application.

Figure 9 shows the results from deep-learning inference benchmarks of some selected Raspberry models and laptop or desktop computer with or without the Intel NCS2 module attached. As we can see that with the Intel NCS2, the performance of all Raspberry Pi models increases rapidly, especially for Raspberry Pi 4 with high speed CPU and USB3.0 port supported. The Intel NCS2 also improves the performance of other hardware platform, such as laptop or desktop computers which do not have a graphic card inside. However, the upper bound benchmark of the Intel NCS2 is about 20 FPS for this experiment.

Table 3 shows the full inference benchmarks and hardware utilization of some selected platforms when running the object detection algorithm. From these results, we can see that the Intel NCS2 not only improves the inference benchmark but also reduces the CPU usage when running deep neural network applications. This allows IoT-enabled devices to adapt well to a wide range of IoT applications.

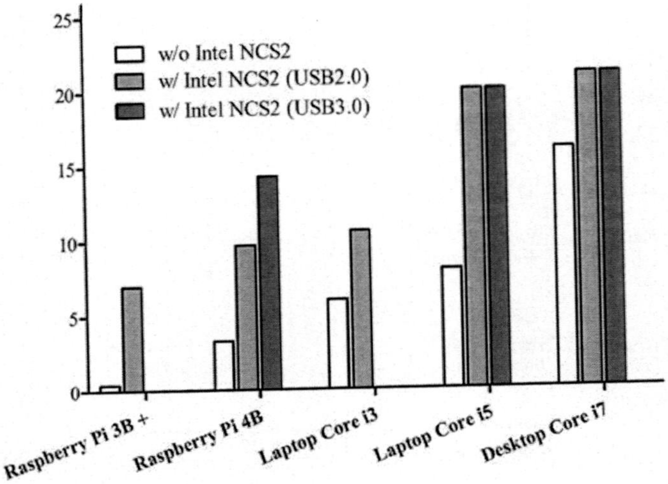

Fig. 9 The comparison of inference benchmarks of some selected platforms

Table 3 Inference benchmarks of some selected platform

Platforms	FPS	Memory usage	CPU usage
Raspberry Pi Zero	0.03	N/A	N/A
Raspberry Pi 3B +	0.3–0.5	83 MB	342%[a]
Raspberry Pi 4B	3.1–3.3	78 MB	343%[a]
Laptop Core i3	5.1–5.9	103 MB	68%
Laptop Core i5	7.3–8.1	100 MB	77%
Desktop Core i7	14.1–16.3	170 MB	60%
Raspberry Pi Zero with Intel NCS2	DNR	DNR	DNR
Raspberry Pi 3B+ with Intel NCS2	6.5–7.5	80 MB	106%[a]
Raspberry Pi 4B with Intel NCS2 (USB 2.0)	8.8–9.7	103 MB	148%[a]
Raspberry Pi 4B with Intel NCS2 (USB 3.0)	13.1–14.3	104 MB	167%[a]
Laptop Core i3 with Intel NCS2 (USB 2.0)	9.4–10.6	107 MB	9%
Laptop Core i5 with Intel NCS2 (USB 2.0)	18.3–20.1	146 MB	7%
Laptop Core i5 with Intel NCS2 (USB 3.0)	18.9–20.2	140 MB	7%
Desktop Core i7 with Intel NCS2 (USB 2.0)	19.8–21.3	196 MB	3%
Desktop Core i7 with Intel NCS2(USB 3.0)	19.5–21.2	193 MB	3%

[a]The total CPU usage of multi-core platform in Linux-based system

DNR (did not run) or N/A: The results occurred due to limited memory capacity or hardware/software limitations

Fig. 10 The mobile robot prototype

3.2 Implementation Results

3.2.1 Mobile Robot Prototype

Figure 10 presents the prototype of our proposed mobile robot. We deploy this robot to the target environment and perform the experiment for functionality testing. We can control the robot motion and pan/tilt camera for better video or photo capturing. We can also get data from sensors to justify the environment.

3.2.2 Robot Control and Management

When accessing to the system, the admin team can perform their operation with full rights. At the Robot Control page (Fig. 11), we can see all collected data such as live streaming, temperature, humidity, MQ2 level, or distance to obstacle. The alarm message can be displayed when the monitoring data reach the threshold value.

At the Data page, all collected information from the target environment can be observed in real time. Figure 12 shows the real-time monitoring data in graphs.

The detailed monitoring data can also be displayed as shown in Fig. 13.

When working in the target environment, the robot can take some photos or record videos which are very helpful to the admin team in the future. An example of recorded videos is presented in Fig. 14.

Fig. 11 The *Robot Control* page

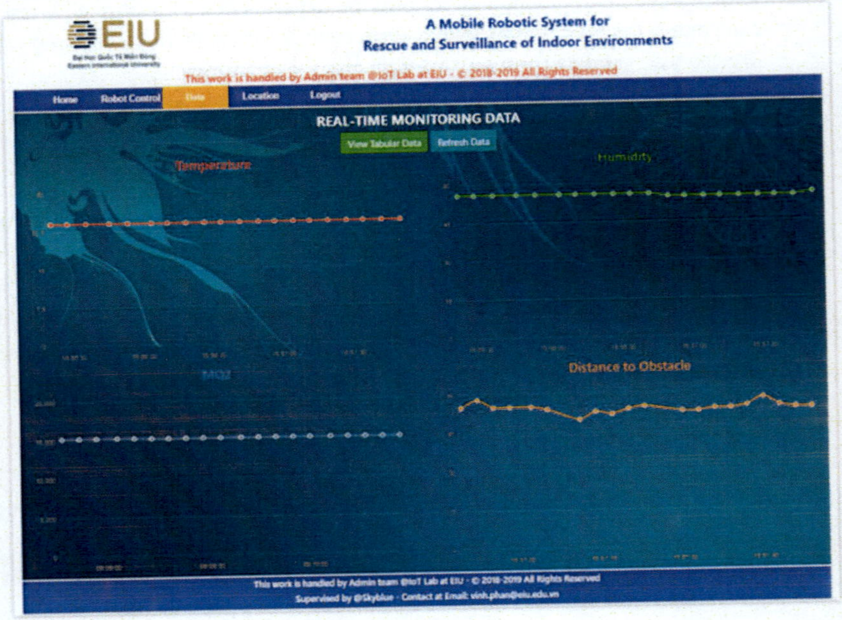

Fig. 12 The *Data* page: monitoring data in graphs

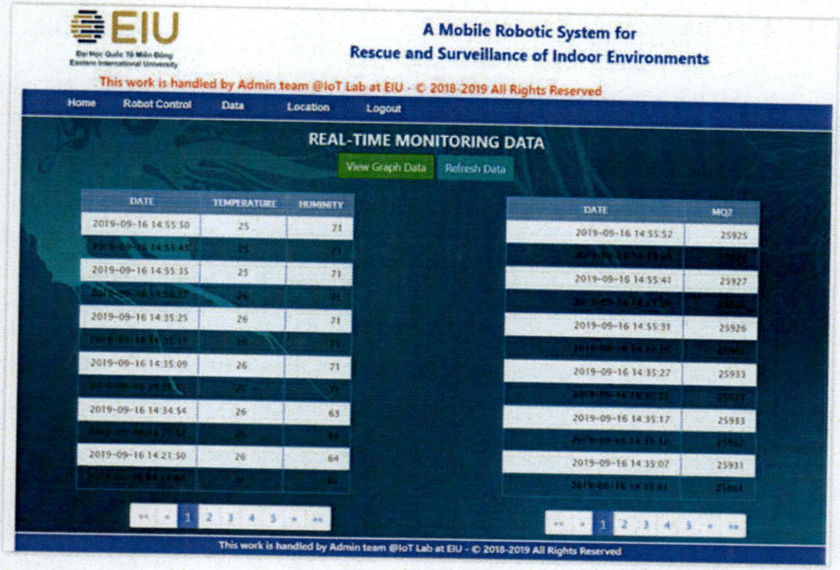

Fig. 13 The Data page: monitoring data in tabular format

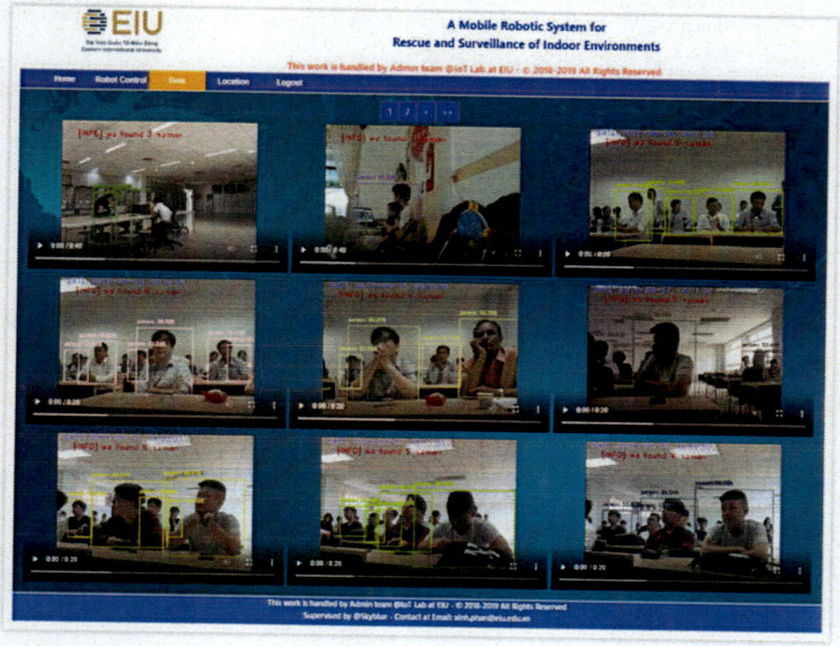

Fig. 14 The Data page: recorded videos with object detection

4 Concluding Remarks

In this paper, we have already applied the newest technology in IoT and computer vision to design and implement a mobile robotic system which collects the ambient data and then analyzes the collected data for making the right actions in a timely manner. The proposed system is potentially helpful in providing the correct information of target environment to the rescue team. Besides, it can detect the objects, especially human being in the given area, which helps to save the human lives quickly and efficiently in case of abnormal or dangerous situation. However, to help our proposed system more practically and efficiently, some necessary features need to be improved as follows. The indoor positioning system needs to be implemented to get the correct position of the collected data (e.g., the real-time and correct position when detecting human beings or emission of poisonous gas in the target area). The prototype hardware model is also redesigned to make the smaller and compact one which can be applied in the practical scenario.

References

1. Di Paola, D., Milella, A., Cicirelli, G., Distante, A.: An autonomous mobile robotic system for surveillance of indoor environments. Int. J. Adv. Robot. Syst. **7**, pp. 19–26 (2010)
2. Budiharto, W.: Intelligent surveillance robot with obstacle avoidance capabilities using neural network. Comput. Intell. Neurosci. **2015**, pp. 1–5 (2015)
3. Shah, M.S., Borole, P.B.: Surveillance and rescue robot using Android smartphone and the Internet. In: International Conference on Communication and Signal Processing (ICCSP), pp. 1526–1530 (2016)
4. Azeta, J., Bolu, C.A., Hinvi, D., Abioye, A.A., Boyo, H., Anakhu, P., Onwordi, P.: An android based mobile robot for monitoring and surveillance. Proc Manuf. **35**, 1129–1134 (2019)
5. Girshick, R., Donahue, J., Darrell, T., Malik, J.: Rich feature hierarchies for accurate object detection and semantic segmentation. In: IEEE Conference on Computer Vision and Pattern Recognition, pp. 580–587 (2014)
6. Girshick, R.: Fast R-CNN. In: IEEE International Conference on Computer Vision (ICCV), pp. 1440–1448 (2015)
7. Cai, Z., Fan, Q., Feris, R., Vasconcelos, N.: A unified multi-scale deep convolutional neural network for fast object detection, In: European Conference on Computer Vision, pp. 354–370 (2016)
8. Ren, S., He, K., Girshick, R., Sun, J.: Faster R-CNN: towards real-time object detection with region proposal networks. IEEE Transactions on Pattern Analysis and Machine Intelligence, pp. 1137–1149 (2017)
9. Redmon, J., Divvala, S., Girshick, R., Farhadi, A.: You only look once: unified, real-time object detection. In: IEEE Conference on Computer Vision and Pattern Recognition (CVPR), pp. 779–788 (2016)
10. Sandler, M., Howard, A., Zhu, M., Zhmoginov, A., Chen, L.: MobileNetV2: inverted residuals and linear bottlenecks. In: IEEE/CVF Conference on Computer Vision and Pattern Recognition, pp. 4510–4520 (2018)
11. https://socialcompare.com/en/comparison/raspberrypi-models-comparison

Study on the Learning in Intelligent Control Using Neural Networks Based on Back-Propagation and Differential Evolution

Shenglin Mu, Satoru Shibata, Huimin Lu, Tomonori Yamamoto, Shota Nakashima, and Kanya Tanaka

Abstract In order to obtain good control performance of ultrasonic motors in real applications, a study on the learning in intelligent control using neural networks (NN) based on differential evolution (DE) is reported in this chapter. To overcome the problems of characteristic variation and nonlinearity, an intelligent PID controller combined with DE type NN is studied. In the proposed method, an NN controller is designed for estimating the variation of PID gains, adjusting the control performance in PID controller to minimize the error. The learning of NN is implemented by DE in the update of the NN's weights. By employing the proposed method, the characteristic changes and nonlinearity of USM can be compensated effectively. The effectiveness of the method is confirmed by experimental results.

Keywords Intelligent control · Neural network · Differential evolution · PID control · Ultrasonic motor

S. Mu (✉) · S. Shibata · T. Yamamoto
Graduate School of Science and Engineering, Ehime University, Matsuyama, Ehime, Japan
e-mail: mu.shenglin.du@ehime-u.ac.jp; shibata.satoru.mg@ehime-u.ac.jp; yamamoto.tomonori.mh@ehime-u.ac.jp

H. Lu
Faculty of Engineering, Department of Mechanical and Control Engineering, Kyushu Institute of Technology, Kitakyushu-shi, Fukuoka, Japan
e-mail: riku@cntl.kyutech.ac.jp

S. Nakashima
Graduate School of Sciences and Technology for Innovation, Yamaguchi University, Ube, Yamaguchi, Japan
e-mail: s-naka@yamaguchi-u.ac.jp

K. Tanaka
School of Science and Technology, Meiji University, Kawasaki, Kanagawa, Japan
e-mail: kanya728@meiji.ac.jp

S. Mu et al. (eds.), *4th EAI International Conference on Robotic Sensor Networks*, EAI/Springer Innovations in Communication and Computing,
https://doi.org/10.1007/978-3-030-70451-3_2

1 Introduction

Along with the development of computer science and electronic engineering, the advancement of artificial intelligent technologies have been progressing in a great pace in recent years. The most representative intelligent scheme Neural Networks (NNs) have been introduced to many industrial fields, and various of applications in our lives. Owing to the NNs' excellent features, they are effectively applied in image/speech processing and classification, emotion recognitions and so on. In the applications of NNs, the most classic approach for the learning in them is the Back-Propagation (BP) method, which is a gradient decent algorithm to obtain the weights and biases in NNs. However, owing to the mechanism using cost function in the BP learning, the differential information is necessary in the applications. For most industrial applications, the differential information of the objectives are difficult to be obtained [1].

Meanwhile, the development of evolutionary computation attracts a lot of attentions in recent years, owing to their excellent features in complex optimization problems. They are considered as good solutions to solve the limitations in the learning of NNs. Genetic algorithm (GA), which is constructed according to the evolutionary mechanisms of natural selection, is the most well-regarded algorithm in the evolutionary computation [2]. Since GAs are with fine searching ability, difficult to be trapped in local minima, and can be applied without considering the differential information in the learning procedure, they have been applied in NN for many applications [3]. However, because of the complexity in the encoding manipulations, and the evolution mechanism in GAs, more effective algorithms with simple manipulations and high efficiency are expected.

In previous research, the NN is proposed for the position control of ultrasonic motor (USM). The traditional approach of NN is applied for the position control of USMs [4]. The BP type NN is also constructed for the position control of USM with speed compensation [5]. The proposed method was confirmed effective and easy to be applied in the control of USM. To solve the problem of Jacobian estimation, we introduced the particle swarm optimization (PSO) algorithms for the NN type PID controller in the control of USM [6, 7]. Owing to the excellent features of PSO in optimization both in continuous and discrete problems [8], the swarm intelligence algorithm was introduced to the proposed scheme, and confirmed effective in obtaining high accuracy in position and speed control of USM. However, there are some concerns about the local minimum problem and ability of re-convergence in PSO for the control process. Therefore, in this study, the algorithm of differential evolution (DE) is introduced in an intelligent control application using NN. According to the study from S. Das et al. [9], DE was proofed being with as good convergence characteristics as PSO. Meanwhile, the hybridizations of both the algorithms are excellent comparing with other soft computing tools. In the study of [10], DE was investigated as a global optimization algorithm for feed-forward NN. The good features of DE, such as no learning rate setting requirement and offering high quality solutions were confirmed. Comparing with the traditional GAs, DE algorithms are with superior features such as simpler construction, high efficiency, and excellent convergence features. In the NN scheme, the DE algorithm is applied to update the weight for learning, without considering the differential limitation as in BP algorithms, or the encoding manipulations as in GAs. The proposed DE type NN is applied to an intelligent PID control scheme for USM to confirm the effectiveness. The intelligent automation with the proposed method is expected to contribute in the medical and welfare applications utilizing the excellent features of USMs [11].

The chapter is organized as follows. In Sect. 2, the general introduction about NN, DE, and the proposed DE type NN scheme is given. In Sect. 3, the simulation study of the proposed intelligent scheme using the DE type NN is introduced. Section 4 gives an experimental study of the proposed method on an existent USM servo system. The conclusions are given in Sect. 5.

2 Proposed Intelligent Scheme

2.1 Neural Network

NNs are machine learning methods that imitate the mechanism of the human brain. Owing to the excellent characteristics of them, they have been studied and applied in various kinds of fields as the most representative of artificial intelligence [12]. Figure 1 shows an example of an NN consisting of three layers of neurons. $input_i$ is the input signal and the neurons contained the signals consist the first layer. w_{ij} is the coupling weight from the first layer to the second layer (hidden layer), H_j is the signal activated according to the sum of the weighted input signals. w_{ij} is the coupling weight from the second layer to the third layer (output layer). O_m is the output signal activated according to the sum of the weighted input signals

Fig. 1 Basic scheme of NN

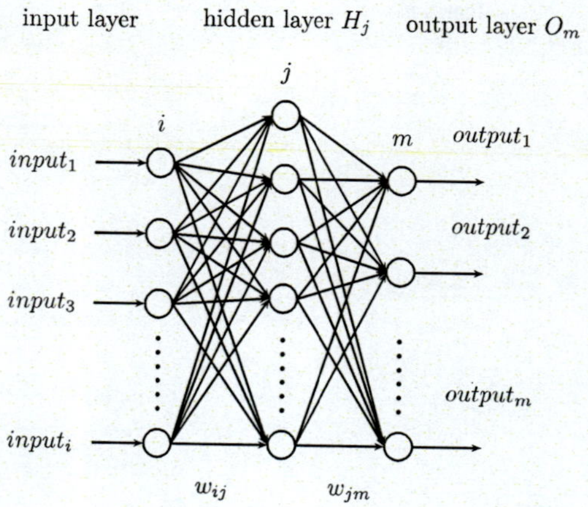

from hidden layer. Theoretically, the weights in general schemes are initialized with random values. However, it is necessary to adjust it to a value suitable for the problem to be dealt with. The BP algorithm is often used for optimization. However, some limitations in traditional methods prevent the BP type NN from wider applications. Therefore, in this study, the DE algorithm is introduced to the learning of NN.

2.2 *Differential Evolution*

In this study, we use the DE algorithm as an optimization method of the weighting coefficient of the NN. It does not need to be differentiable because it is one of the methods and does not use the gradient of the problem to be optimized. Due to its characteristics, the DE algorithm can be applied even when the problem is discontinuous, or time-varying. The general procedure in DE can be depicted by the flowchart shown in Fig. 2.

2.2.1 Randomly Initialization

Basically, DE algorithm is a global optimization algorithm for obtaining optimal solutions in a D dimensional space. The initialization of the algorithm is usually implemented by random real numbers consisting a vector in the D dimension. The randomly initialized vectors are considered as possible solutions, which can be represented as

$$X_i = [x_1, x_2, \ldots, x_D] \tag{1}$$

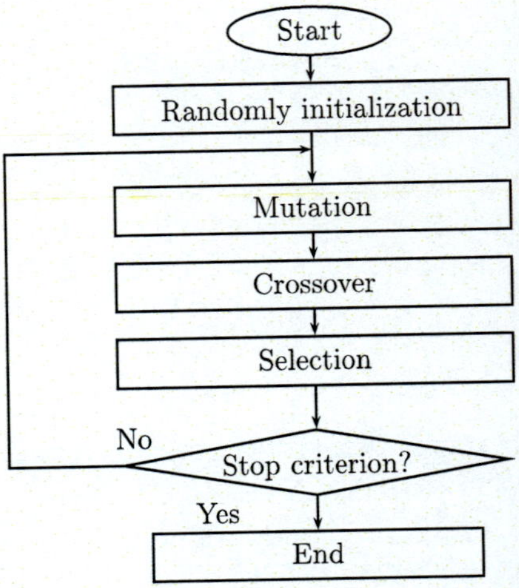

Fig. 2 Flowchar of DE algorithm

If the searching space can be constrained by

$$X_{i,Min} = [x_{1,Min}, x_{2,Min}, \ldots, x_{D,Min}] \tag{2}$$

$$X_{i,Max} = [x_{1,Max}, x_{2,Max}, \ldots, x_{D,Max}] \tag{3}$$

Then, the elements in the vector can be randomly initialized by

$$x_i = x_{i,Min} + rand(0,1) \cdot (x_{i,Max} - x_{i,Min}) \tag{4}$$

where $rand(0, 1)$ is the random number within the range of (0, 1).

2.2.2 Mutation

Following the random initialization, the mutation according to target vector is implemented. The donor vector v_i can be generated as linear combination of some vectors in current searching generation selected according to some strategies. There are some strategies introduced in [13]. In this study, we focus on four strategies as follows:

- DE/rand/1/bin

$$v_i = x_{r1} + F \cdot (x_{r2} - x_{r3}) \tag{5}$$

- DE/rand/2/bin

$$v_i = x_{r1} + F \cdot (x_{r2} - x_{r3}) + F \cdot (x_{r4} - x_{r5}) \tag{6}$$

- DE/current-to/1/bin

$$v_i = x_i + F \cdot (x_{r2} - x_{r3}) \tag{7}$$

- DE/best/1/bin

$$v_i = x_{best} + F \cdot (x_{r2} - x_{r3}) \tag{8}$$

$r1$, $r2$, $r3$, $r4$, $r5$ are the randomly chosen integers with different value from each other and i, within the range of [1, N(number of searching points)].

2.2.3 Crossover

In crossover, the donor vector is applied to be combined with the target vector. There are mainly two kinds of crossover methods (exponential and binomial) applied in DE algorithms. In this study, we employ the binomial crossover, which applies comparison between a random number generated in (0, 1) and a parameter of crossover probability C_r to decide the crossover scheme. The trial vector of U_i can be generated, and the elements in it can be calculated as

$$u_{i,j} = \begin{cases} v_{i,j} & \text{if} rand(0,1) \leq C_r, \quad \text{or} \quad j = j_{rand} \\ x_{i,j} & \text{otherwise.} \end{cases} \tag{9}$$

where *rand*(0, 1) is a real number random number in [0, 1], and j_{rand} is an integer random number in 1, ..., D.

2.2.4 Selection

In the selection procedure, the generated target vector is decided to be a survival or not. The decision is made according to the evaluation of objective/evaluation function $f(\cdot)$. The procedure can be expressed as

$$X_i = \begin{cases} U_i, & \text{if} f(U_i) \leq f(X_i) \\ X_i, & \text{if} f(U_i) > f(X_i) \end{cases} \tag{10}$$

2.3 *Proposed DE Type NN Scheme*

In this study, the DE algorithm is introduced to update the weights in the learning of NN. The method can be constructed with simple structure, easy to be applied, without considering the differential information.

In the proposed scheme, the vector of DE is designed as the real value of weights in NN. The target vector in DE can be expressed as

$$W = [x_{00}, x_{01}, \ldots, x_{ij}, \ldots, x_{jm}] \tag{11}$$

The weights in the NNs are updated by the mutation, crossover, and selection procedures in the DE algorithm to obtain the optimal solutions in NN's learning.

3 Simulation Study

To confirm the effectiveness of the proposed scheme, the simulation study was implemented employing the Rosenbrock function for optimization using NN. The Rosenbrock function is a convex function with two input values, defined as x and y as follows.

$$f(x, y) = (1 - x)^2 + 100(y - x^2)^2 \tag{12}$$

The function can be plotted as in Fig. 3. In the study, the NN with the topologic structure shown in Fig. 4 was introduced to find the minimum of the Rosenbrock function. The NN is constructed to find the optimal (x, y) to minimize the Rosenbrock function.

The weights in the NN is designed to be updated by the DE introduced in Sect. 2. The weights are initialized by the random real number within the range of (−1.0, 1.0). The mutation strategies were investigated in the simulation to confirm the characteristics in DE. The parameter F which is the scaling factor is set as 0.5. The crossover probability C_r is set as 0.8. The number of searching unit in DE is set as 10. To investigate the effectiveness of the proposed scheme comparing with the conventional method, we also implement the NN's learning using traditional BP algorithm. The weights of NN in BP is updated as follows. The weights between the output layer and the hidden layer can be updated as

$$w_{jm} = w_{jm} - \eta \frac{\partial f}{\partial w_{jm}} \tag{13}$$

where η is the learning rate in NN, and w_{jm} is the weights between the hidden layer and the output layer. Then, the weights w_{ij} between the input layer and the hidden layer can be updated as

$$w_{ij} = w_{ij} - \eta \frac{\partial f}{\partial O_m} \frac{\partial O_m}{\partial H_j} \frac{\partial H_j}{\partial w_{jm}} \tag{14}$$

By applying both the methods of DE type NN and BP type NN, the NNs are trained to converge to the optimal values with proper weights. Outputs of the BP and DE type NNs are shown in Fig. 5. The figure shows that the outputs of both NNs converged to the optimal (x,y) to minimize the value of the function in hundreds of iterations. Comparing with the convergence of the BP type NN, the DE type NN showed better convergence characteristics. The errors of the minimum value and the

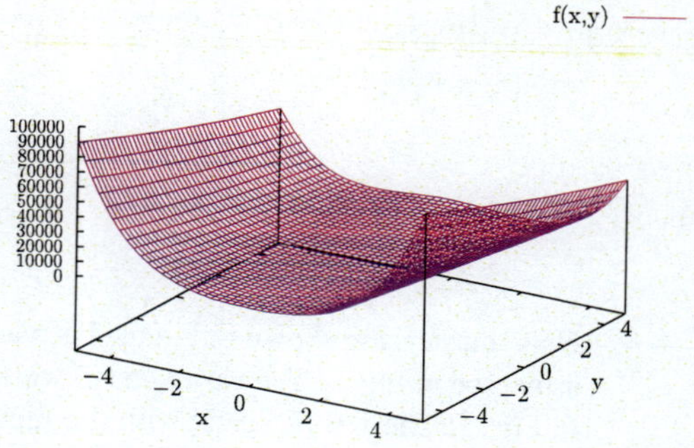

Fig. 3 The plot of the Rosenbrock function

Fig. 4 Structure of DE-NN for simulation

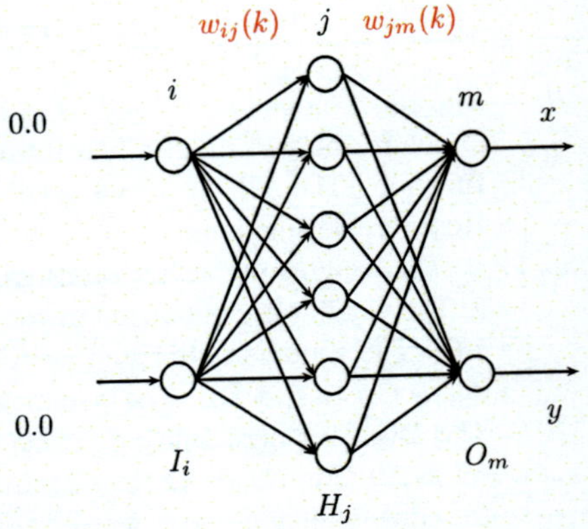

Fig. 5 Outputs of the BP and DE type NNs

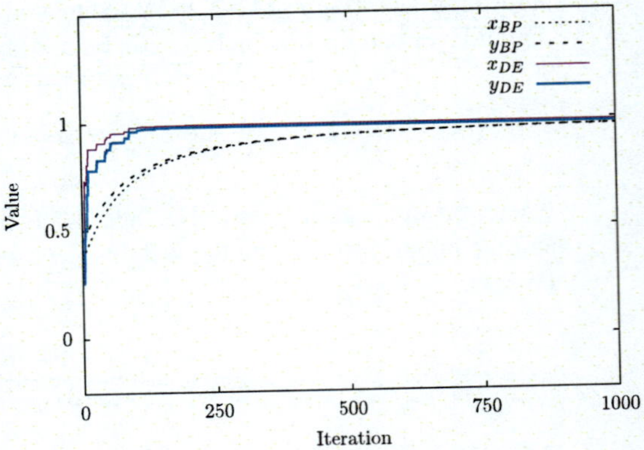

Fig. 6 Errors of the BP and DE type NNs

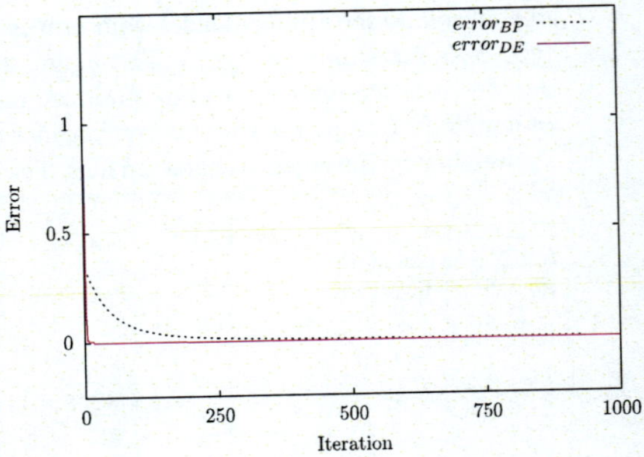

NNs' outputs are shown in Fig. 6. The variation of the errors shows the same trends as we see in Fig. 5. The variations of weights in the BP and DE type NNs are shown in Fig. 7. The DE type NN with the mutation strategy of DE/best/1/bin. It can be

Fig. 7 Weight convergence of the BP and DE type NNs

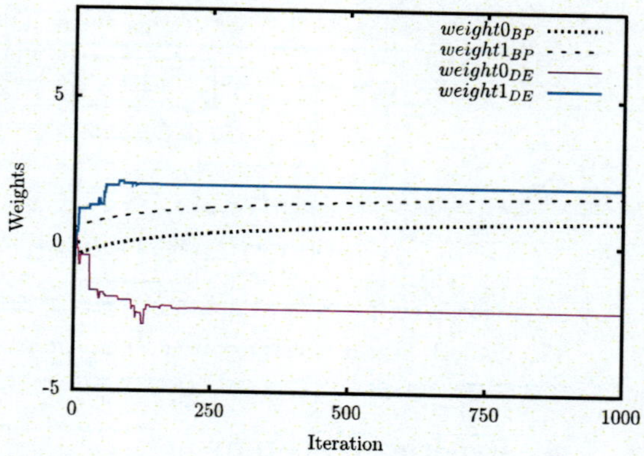

Fig. 8 Convergence of DE with different mutation strategies

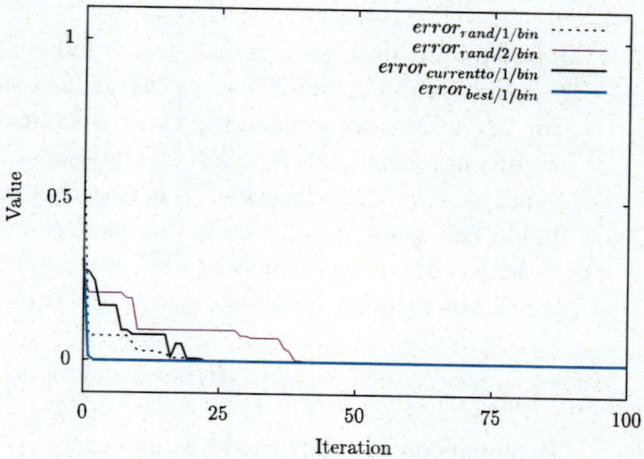

seen that the DE update the NN's weights in iterations. The weights can converge to certain constant values in short time. The values of weights converged quick in around 100 iterations. Comparing the proposed DE type NN with the traditional BP type NN, the proposed scheme was with an excellent convergence features.

To confirm the conference characteristics of DE with different mutation strategies, the simulation study was implemented in the optimization of Rosenbrock function using the proposed NN. The DEs with the mutation strategies of DE/best/1/bin, DE/current to/1/bin, DE/rand/1/bin and DE/rand/2/bin were investigated [13]. The variation of the evaluation value was studied as Fig. 8 shows. According to the simulation results, it is clear to see that among the DEs with the strategies, the DE/best/1/bin strategy converged to quite low evaluation value in only a few iterations. The DE/best/1/bin strategy is with great convergence characteristics to approach the optimal. Therefore, the DE/best/1/bin is considered as the best solution for experimental study.

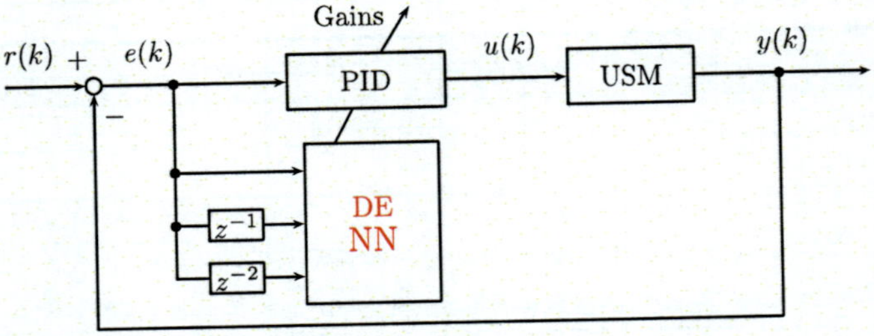

Fig. 9 Block diagram of the proposed intelligent method using DE type NN

4 Experimental Study

Based on previous research in [7], an intelligent control method using PID control combined with DE type NN is applied as the controller for USM as shown in Fig. 9. The NN scheme we propose is a fixed structure as we see in Fig. 10. In the proposed intelligent scheme, DE type NN is employed for updating the PID gains. The error signal of $[e(k), e(k-1), e(k-2)]$ is employed as input. The scheme is designed for tuning PID gains automatically to minimize the error in USM control. The output of the NN unit is the variation of PID gains $[\Delta K_P(k), \Delta K_I(k), \Delta K_D(k)]$. The output of neurons in hidden-layer is expressed as $H_j(k)$, which can be estimated by Eq. 15.

$$H_j(k) = f_s(w_{ij}(k) \cdot I_i(k)) \tag{15}$$

Then, the output of NN can be estimated as

$$\Delta K_{P,I,D}(k) = f_s(w_{jm}(k) \cdot H_j(k)) \tag{16}$$

where $f_s(x)$ is the sigmoid function as shown in Eq. 17.

$$f_s = \frac{1}{1+e^{-x}} - 0.5 \tag{17}$$

The weights between three layers, expressed by $w_{ij}(k)$ and $w_{jm}(k)$, are updated by BP and DE algorithms. Then, the output of PID controller $u(k)$ can be calculated as

$$\begin{aligned} u(k) = u(k-1) + (K_P(k) + K_I(k) + K_D(k))e(k) \\ - (K_P(k) + 2K_D(k))e(k-1) + K_D(k)e(k-2) \end{aligned} \tag{18}$$

In our experimental study, the sin wave was applied as the input signal. The response of the proposed method is shown in Fig. 11. There was no visible error in the response of sin wave input. Figure 12 shows the variation of the PID gains updated by the proposed scheme using DE type NN. The gains oscillated in the

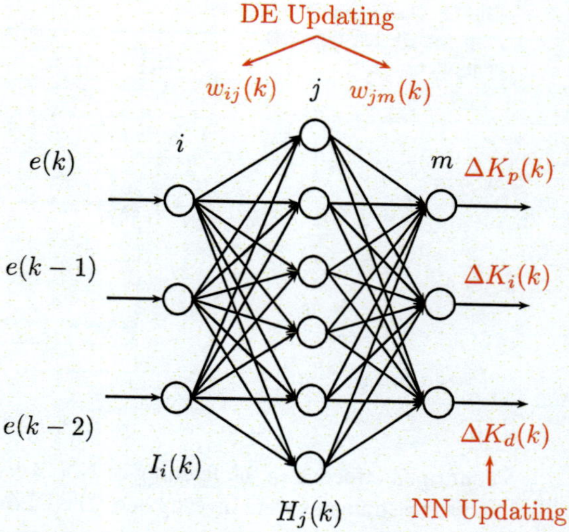

Fig. 10 The topologic structure of NN controller

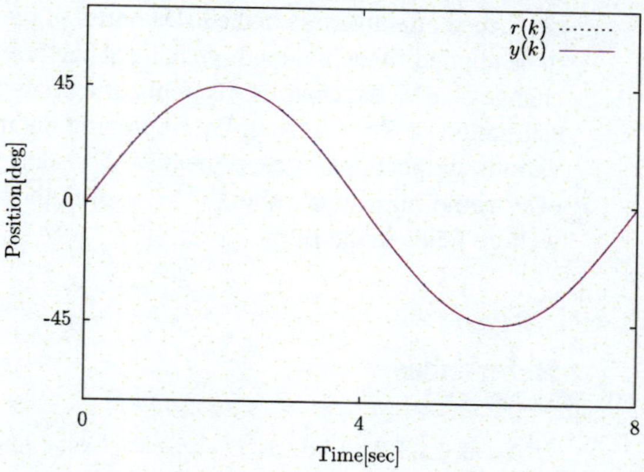

Fig. 11 Response for sin wave signal input using DE type NN

initial phase in a short period. Then, they converged to certain values quickly. The convergence and accuracy of the proposed method is confirmed according to the experimental results.

5 Conclusions

In this chapter, an intelligent PID control scheme using the DE type NN is proposed. In the proposed scheme, an NN method is employed for optimizing the gains in PID controller for the control of ultrasonic motor. The weights of the NN are designed to be updated by DE algorithm. By employing simulation study, the DE algorithm was

Fig. 12 Gain variation by the proposed DE-NN type PID controller

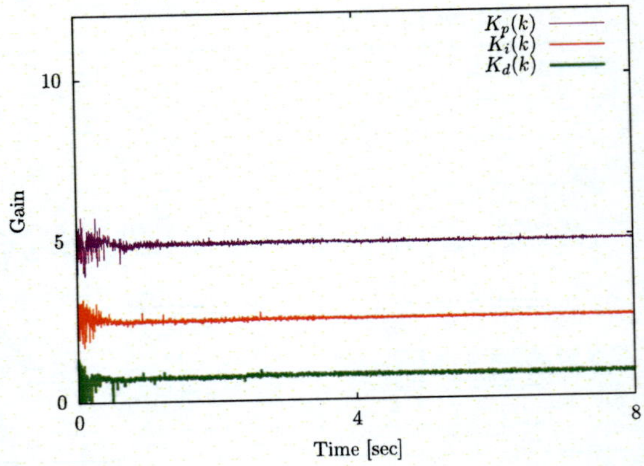

confirmed effective in the learning of NN. According to the simulation results, the mutation strategies were investigated. The DE/best/1/bin was the most effectiveness in the optimization process. By the experimental study, the DE type NN was confirmed effectiveness in the PID control of the USM. The proposed method is able to be applied for achieving high control performance in compensating characteristic changes USM. Meanwhile, its application is without any consideration of Jacobian estimation in NN's leering. By employing the method, USMs are expected more in various applications, such as the meal-assistance robots [14], human support robots, even some humanoid robots [15], especially for the applications in medical and welfare fields in the future.

References

1. Tsoukalas, L.H., Uhrig, R.E.: Fuzzy and Neural Approaches in Engineering. Wiley, New York (1997)
2. Srinivas, M., Patnaik, L.M.: Genetic algorithms: a survey. Computer **27**(6), 17–26 (1994)
3. Wang, L.: A hybrid genetic algorithm – neural network strategy for simulation optimization. Appl. Math. Comput. **170**(2), 1329–1343 (2005)
4. Senjyu, T., Miyazato, H., Uezato, K.: Position control of ultrasonic motor using neural network (Japanese). Trans. Inst. Electr. Eng. Jpn. D **116**, 1059–1066 (1996)
5. Oka, M., Tanaka, K., Uchibori, A., Naganawa, A., Morioka, H., Wakasa, Y.: Precise position control of the ultrasonic motor using the speed compensation type NN controller (Japanese). J. Jpn. Soc. Appl. Electromagn. Mech. (JSAEM) **70**(6), 1715–1721 (2004)
6. Mu, S., Kanya, T.: , Intelligent IMC-PID control using PSO for ultrasonic motor. Int. J. Eng. Innov. Manag. **1**(1), 69–76 (2011)
7. Mu, S., Tanaka, K., Nakashima, S., Alrijadjis, D.: Real-time PID controller using neural network combined with PSO for ultrasonic motor. ICIC Exp. Lett. **8**(11), 2993–2999 (2014)
8. Jordehi, A.R., Jasni, J.: Particle swarm optimisation for discrete optimisation problems: a review. Artif. Intell. Rev. **43**, 243–258 (2015)

9. Das, S., Abraham, A., Konar, A.: Particle swarm optimization and differential evolution algorithms: technical analysis, applications and hybridization perspectives. Stud. Comput. Intell. **116**, 1–38 (2008)
10. Ilonen, J., Kamarainen, J., Lampinena, J.: Differential evolution training algorithm for feed-forward neural networks. Neural Process. Lett. **17**, 93–105 (2003)
11. C. Zhao, Ultrasonic Motor - Technologies and Applications. Science Press/Springer Beijing/Berlin (2011)
12. Basheer, I.A., Hajmeer, M.: Artificial neural networks: fundamentals, computing, design, and application. J. Microbiol. Methods **43**, 3–31 (2000)
13. Islam, S.M., Das, S., Ghosh, S., Roy, S., Suganthan, P.N.: An adaptive differential evolution algorithm with novel mutation and crossover strategies for global numerical optimization. IEEE Trans. Syst. Man Cybern. B (Cybernetics) **42**(2), 482–500 (2012)
14. Tanaka, K., Kodani, K., Oka, M., Nishimura, Y., Farida, F.A., Mu, S.: Meal assistance robot with ultrasonic motors. Int. J. AEM **36**, 177–181 (2011)
15. Zhou, C., Wang, X., Li, Z., Tsagarakis, N.: Overview of gait synthesis for the humanoid COMAN. J. Bionic Eng. **14**(1), 15–25 (2017)

Word Sense Disambiguation Based on Graph and Knowledge Base

Fanqing Meng

Abstract Word sense disambiguation determines the specific meaning of ambiguous words according to their specific context, a basic research in the field of natural language processing, and has a direct impact on the upper application of machine translation, information retrieval, text classification, and sentiment analysis. This paper proposes a word sense disambiguation method based on graph and knowledge base, for the insufficient utilization problem of existing knowledge base. It utilizes dependency parsing to obtain contextual knowledge and processes the examples in lexical knowledge base, which have a good ability of sense distinction, to construct disambiguation graph. Then, the disambiguation can be done by combining dependency disambiguation graph and contextual disambiguation graph. Experiments on the dataset of SemEval-2007 task#5 show that the disambiguation accuracy is 0.471, which is better than the mentioned methods.

Keywords Word sense disambiguation · Graph · Dependency parsing

As a basic research of natural language processing, word sense disambiguation (WSD) determines the specific meaning of ambiguous words according to their specific context, which has a direct impact on high-level applications, such as machine translation, information extraction, information retrieval, text classification, and emotion analysis [1–4]. Whether it is Chinese or English and other Western languages, the phenomenon of polysemy is widespread. According to statistics [5, 6], the frequency of ambiguous words in Chinese corpus is about 42% and the frequency of ambiguous words in English corpus is 30–43%. Obviously, ambiguous words are frequent in natural language text, which makes the performance of WSD seriously affect the effect of upper natural language processing applications. In recent years, the technique of graph-based WSD has been widely concerned

F. Meng (✉)
School of Information Science and Engineering, Zaozhuang University, Zaozhuang, China
e-mail: mengfanqing@uzz.edu.cn

S. Mu et al. (eds.), *4th EAI International Conference on Robotic Sensor Networks*, EAI/Springer Innovations in Communication and Computing,
https://doi.org/10.1007/978-3-030-70451-3_3

by researchers. This paper attempts to use HowNet and the Chinese Semantic Dictionary as the knowledge base for Chinese WSD tasks.

1 Related Work

Most WSD methods based on graph are inspired by lexical chain. The lexical chain means some semantic related words in a given text that form a sequence and the words that are linked together through lexical semantic relationships. Galley et al. put forward a method of WSD based on lexical chain [7], which is mainly divided into two steps: first, when constructing the graph model, senses of the target ambiguous word are added to the graph, the words in the text are processed sequentially, and the senses of each word are compared with the words that have been processed. If there is some semantic relationships between them, then the relationship is used as a link edge in the graph, and the weight is arranged according to the semantic relation and the distance between the words. Second, in the disambiguation stage, the weight of each word sense and context-linked word node is computed, and the word sense with the largest weight is selected as the correct word sense of the target word. This method achieves 62.1% disambiguation accuracy on the noun dataset of SemCor.

Mihalcea proposes a disambiguation method based on the PageRank algorithm [8], which takes the senses of each word in the text as the vertex and constructs the disambiguation graph by using the semantic relation between the senses of words. In addition to the semantic relationships in WordNet, it includes a coordinating relation to link the sense of the same upper concept. In this paper, the PageRank algorithm is used to calculate the importance of each sense node iteratively, and the maximum score of the senses of ambiguous words is chosen as its final sense. Navigli et al. propose the structured semantic interconnect algorithm (SSI) [9]. With the help of knowledge resources such as WordNet, the algorithm structurally expresses the senses of contextual words; it constructs grammar rules by manual means to describe the interconnection patterns in semantic concept mappings; it performs disambiguation iteratively, giving priority to ambiguous words with less difficulty. According to the semantic interconnection mode of words that have been determined, the disambiguation words are processed. This method achieves the best results in the international semantic evaluation Senseval-3 and SemEval-2007. Agirre et al. propose the Personalized PageRank algorithm for WSD [10, 11] which is similar in principle to Mihalcea's method [8] and the difference is that it has some improvements in the importance of degree computation, to modify the importance of some word nodes. Nieto Pina et al. present a graph-based WD method for Swedish, and it trains embeddings by using PageRank on the graph of lexical knowledge base (LKB) [12]. Janz et al. propose a WSD method for English and Polish based on Monte Carlo approximation and PageRank [13]. It utilizes the local semantic relatedness by an extensive knowledge base. Scozzafava et al. propose a knowledge-based multilingual WSD system, and it performs WSD using PageRank and syntagmatic information [14]. These methods mainly focus on the English WSD task, but the Chinese semantic computing resources are relatively scarce, which limits the popularization and application of these methods in Chinese WSD tasks.

Yang et al. propose a network graph disambiguation method based on word distance [15]. The method performs word sense disambiguation according to the distance between the vertices of the words in the network graph. The vertices of words closer to ambiguous words are strengthened, and the vertices of distant words are weakened, that is, words with close distance have a strong recommendation effect on ambiguous words, and words with far distance have weak recommendation. Lu et al. proposes a graph model disambiguation method based on domain knowledge [16]. It applies the domain knowledge to the graph model and improves the importance of degree scores of the sense nodes by improving the various graph scoring methods and then selects the correct sense. Although these methods improve the WSD effect, they do not explore and utilize the existing Chinese word sense computing resources.

2 Method

HowNet reveals the relations between concepts and the attributes of concepts [17], such as hypernym, hyponymy, synonymy, and meronymy. The traditional HowNet-based WSD method has the problem of insufficient knowledge utilization of HowNet. It does not fully exploit the disambiguation knowledge existing in HowNet. For example, the example sentences in HowNet have a strong ability to distinguish word sense. By digging deep into the semantic knowledge contained in HowNet, it can be used as the basis for WSD, which will contribute to the improvement of WSD performance.

This paper proposes a WSD method based on graph and knowledge base. It analyzes the ambiguous sentences, obtains the context-dependent tuples, and obtains the dependency graph to construct the context disambiguation graph. In HowNet, the target ambiguous example sentences are obtained, and the dependency parsing is performed to get the dependent tuple set. Combining with context-dependent tuples, construct dependent disambiguation graph, fuse context disambiguation graph and dependent disambiguation graph, use graph scoring algorithm to score the importance of degree of word sense vertices, and select the one which gets the biggest score in the senses of target ambiguous word as the correct word sense.

2.1 HowNet

HowNet is a common-sense knowledge base that Professor Dong spends more than 10 years organizing and constructing [17]. It builds a network of knowledge base based on the relation between the concepts of Chinese and English, and the attributes contained in the concepts. HowNet is mainly composed of a knowledge dictionary, a knowledge management system, and documentation. Among them, the knowledge dictionary is the basic file of HowNet, which is based on words and the concepts represented by words.

The choice of words in HowNet is based on the frequency word list of 400 million words in Chinese corpus, rather than just referring to a specific semantic dictionary; it pays attention to collecting popular and fixed words, but not blindly seeking new ones. Considering the circulation of word sense in modern times, if one word has two senses, one of which is more commonly used and the other is seldom used, only the former sense will be retained. At the same time, it provides the corresponding English word interpretation for the word concept. Taking the sense of a commonly used polysemous word "打" as an example, its specific concept in HowNet is defined as follows:

NO.=023671
W_C=打
G_C=V [da3]
S_C=
E_C=给他~电话, 怎么也~不通, 对不起~错了, 给他~了好几通电话, ~过去了吗, ~通了但没人接, 我给你~了一个晚上
W_E=call
G_E=V
S_E=
E_E=I tried to call you all night
DEF={communicate|交流:instrument={tool|用具:{communicate|交流:instrument = {~}}}}

Among them, "NO." indicates the number of the concept and uniquely identifies a specific word sense; "W_C" indicates the specific lexical form corresponding to the concept; "G_C" indicates part of speech and pinyin; "S_C" indicates conceptual emotional information; "E_C" is the example sentence, which is intended to highlight the distinguishing ability of the word sense rather than to emphasize its ability to interpret it, which is helpful for WSD; "W_E" means the English word corresponding to the concept; "G_E" means part of speech of the English word; "S_E" means the emotional information of the English word; "E_E" means examples of English words; and "DEF" refers to the definition of a concept. Sememe is the smallest semantic unit in HowNet. The concept definition in HowNet is composed of sememe. The first sense in the definition is the primary sense, which represents the main semantic information of the concept.

2.2 The Chinese Semantic Dictionary

The Chinese Semantic Dictionary (CSD) is a semantic knowledge base for natural language processing, and it provides a huge amount of semantic information, which includes more than 66,539 entries, Chinese words and their English counterparts [18]. Examples sentences in CSD can be extracted as a supplement for HowNet.

2.3 *WSD Based on Graph and Knowledge Base*

2.3.1 The Disambiguation Framework

The disambiguation framework of the WSD method based on graph and knowledge base proposed in this paper is shown in Fig. 1. Obtain dependency parsing of ambiguous sentence, obtain context-dependent tuples, get dependency graph, and construct context disambiguation graph; obtain example sentences in HowNet, perform dependency parsing, get dependent tuple sets, and combine context-dependent tuples to construct a dependency disambiguation graph; fuse the context disambiguation graph and the dependency disambiguation graph to perform word sense disambiguation. The specific description is as follows:

1. Obtain dependency parsing of ambiguous sentences, obtain context-dependent tuple sets, and get a context-dependent graph; according to the shortest path length of the context words in the dependency graph from the target ambiguous word, the context words are extracted as context knowledge.
2. Use HowNet to label the senses of context knowledge and target ambiguous word, which may exist in HowNet, and compute the similarity among them.
3. Take the senses of the word in the context knowledge as the vertex and the target ambiguous word as the vertices, the semantic relation between concepts as the edge, the concept similarity as the weight of the edge, to construct the context disambiguation graph.
4. Obtain the example sentences of the senses of the target ambiguous word in HowNet (and CSD), perform dependency parsing to obtain the example sentence-dependent tuple set, and construct the dependency disambiguation graph in combination with the context tuple set.

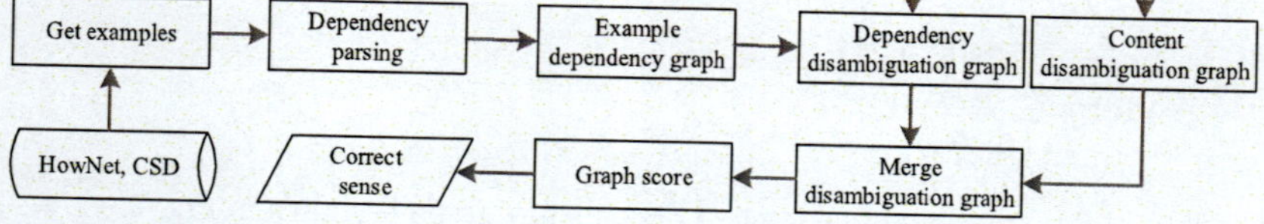

Fig. 1 The disambiguation framework

5. Merge the context disambiguation graph and dependency disambiguation graph, and use graph algorithm to score the importance of degree of each word sense vertex.
6. Select the sense of the ambiguous word which gets the highest score as the correct sense.

2.3.2 Dependency Parsing

The dependency parsing was first proposed by the French linguist L. Tesniere [19]. Dependency parsing analyzes the structure of sentence based on dependent grammatical rules to determine the dependencies between words or phrases and their role in sentence. Dependency parsing results can be expressed in the form of dependency graph. The dependency parser used in this paper is the Stanford CoreNLP natural language toolkit [20], which contains models for syntactic analysis of multiple languages. The model files for Chinese processing are chineseFactored.ser.gz, chinesePCFG.Ser.gz, xinhuaFactored.ser.gz, and xinhuaPCFG.ser.gz. Among them, if there is "Factored" in the name of a model file, that means it contains lexical information, "PCFG" is a faster and smaller template, "xinhua" is trained according to the Xinhua Daily corpus, and "Chinese" also contains Hong Kong and Taiwan corpus. This paper uses the ChineseFactored.ser.gz model file.

In this paper, two tasks are completed by means of dependency parsing. One is to obtain the context dependency graph of the ambiguous sentence and the other is to obtain the example sentence dependency graph corresponding to the sense of the target ambiguous word. The details are as follows.

1. Construct a context dependency graph

The results of the dependency parsing can be expressed in the form of a triple, specifically: dependency (dominant, subordinate). For example, dependency parsing "她生养了一双可爱的儿女。", you can get a context-dependent tuple set as follows:

```
nsubj(生养-2, 她-1)
root(ROOT-0, 生养-2)
asp(生养-2, 了-3)
nummod(双-5, 一-4)
clf(儿女-8, 双-5)
amod(儿女-8, 可-6)
mark(可爱-6, 的-7)
dobj(生养-2, 儿女-8)
punct(生养-2, 。-9)
```

The corresponding context dependency graph is shown in Fig. 2.

Firstly, the ambiguous sentence is analyzed by dependency parsing to obtain a context dependency graph. According to the shortest path length of the context word and the target ambiguous word in the dependency graph, the context knowledge is obtained. The context knowledge here refers to nouns, verbs, adjectives, adverbs, etc. According to the dependency graph, the words with the ambiguous word within a certain shortest path range can be obtained as the context knowledge. As shown in Fig. 2, when the shortest path is set to 1, the context knowledge is obtained as "生养" and "可爱".

2. Build an example sentence dependency graph

Get the example sentences of the ambiguous word "儿女" in HowNet. The first word has five short examples, "一双~, 养育~, 有~的人, 懂事的~, 供应~的学费", using dependency parsing to obtain a set of dependent tuples corresponding to each example sentence. The example "一双~" according to "nummod(儿女-2, 一双-1)"; "养育~" according to "dobj(养育-1, 儿女-2)"; "有~的人" according to "assmod(人-4, 儿女-2), case(儿女-2, 的}-3), dobj(有-1, 人-4)"; "懂事的~" according to "relcl(儿女-3, 懂事-1), mark(懂事-1, 的-2)"; "供应~的学费" according to "assmod(学费-4, 儿女-2), case(儿女-2, 的-3), dobj(供应-1, 费-4)". Further, the dependency graph can be obtained by converting the dependent tuple set, as shown in Fig. 3. The second word sense has two examples "~情长, ~私情", respectively, to analyze using dependency parsing, obtain the dependent tuple set "nsubj(情长-2, 儿女-1)", "nn(私情-2, 儿女-1)", and then get the example dependency graph, as shown in Fig. 4.

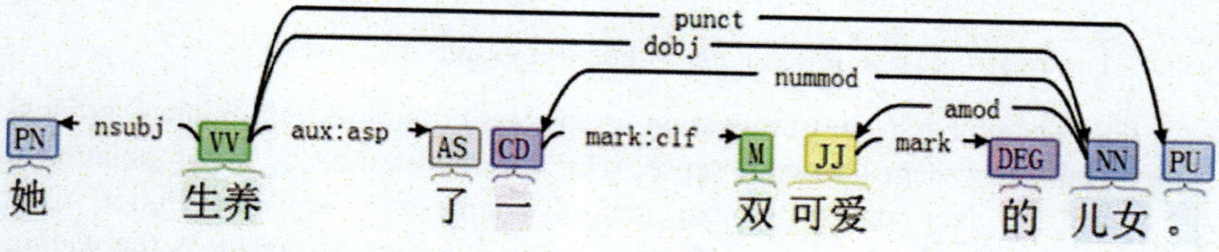

Fig. 2 Context dependency graph

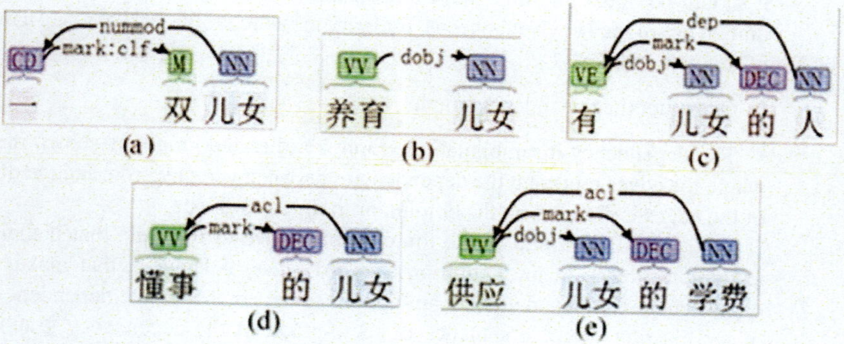

Fig. 3 The example dependency graph of the first sense

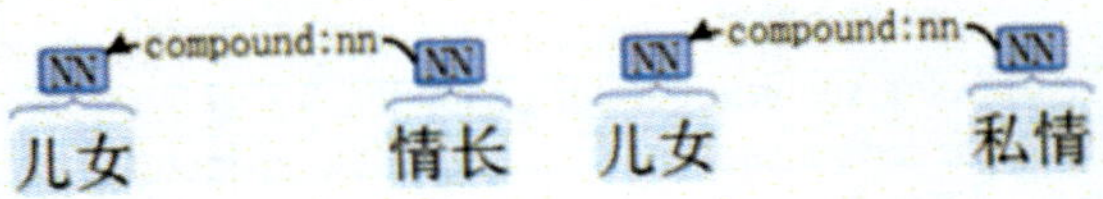

Fig. 4 The example dependency graph of the second sense

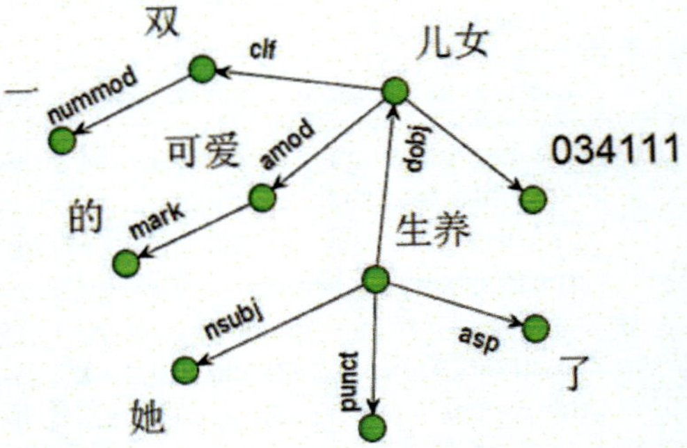

Fig. 5 Dependency disambiguation graph

2.3.3 Construct the Disambiguation Graph

There are two works in this section: one is to construct the context disambiguation graph using the context dependency graph obtained in Sect. 2.3.2 and the other is to use the context dependency graph and the example dependency graph (see Sect. 2.3.2) to construct a dependency disambiguation graph.

1. Construct the context disambiguation graph

   The context disambiguation graph is an undirected graph. Vertices in the graph represent the word sense, which is represented by a combination of word and HowNet concept number, and edge represents semantic relation. Conceptual similarity calculation is performed using HowNet. The result is the weight of the edge. Semantic relation refers to hypernymy, hyponymy, synonymy, etc. In HowNet, it should be noted that this is not refined into a specific semantic relation, but the unified use of the HowNet concept similarity measure toolkit [21] to make a measure of semantic relation.

2. Construct the dependency graph

   The dependency disambiguation graph is a directed graph, as shown in Fig. 5, where the edges represent the dependencies between vertex words, and the direction of the edges is directed by the dominator to the subordinate.

   Next, it will continue to use the example above to illustrate that if there is an example sentence in the definition of word sense of HowNet that makes all the dependency relations match the ambiguous sentence, that is, the dependency graph of the existing example sentence is a sub-graph of the dependency graph of the ambiguous sentence, then the dependency graph of the ambiguous sentence and its sense example sentence will be merged, and the corresponding sense will be labeled in the graph. The ambiguous sentence in the example above is “她生养了一双可爱的儿女。”, and its dependency graph is as shown in Fig. 2; the example sentences of the first sense of the ambiguous word "儿女" are "一双~,养育~, 有~的人, 懂事的~, 供应~的学费". The corresponding dependency graph is as shown in Fig. 3. Obviously, the dependency graph of the example sentence "一双~" is a sub-graph of the dependency graph of the ambiguous sentence (see Fig. 2). At this time, the two are combined and the corresponding word sense information is labeled, specifically the concept number "034111" in HowNet, as shown in Fig. 5.

3 Experiment and Analysis

3.1 Dataset Selection and Evaluation Method

The experimental data set comes from the Chinese and English vocabulary task of international semantic evaluation SemEval-2007, using the standard evaluation method provided by the official, macro average *Pmar*.

3.2 Contrast Method

TorMd This method uses naive Bayes classifier which is an unsupervised method for the University of Toronto to participate in the SemEval-2007 task5 evaluation and won the first place in the evaluation.

Win This method only uses window-based method to select the left and right ambiguous words when selecting context knowledge. Other settings are consistent with the methods proposed in this paper.

Con This method refers to the context disambiguation graph.

CSD This method is the same as HowCSD but does not use HowNet.

HowCSD This method refers to the combination of context disambiguation graph and dependency disambiguation graph for word sense disambiguation.

3.3 Experiment and Analysis

The disambiguation effect of each method is shown in Table 1. It can be seen that the effect of Con method is not more than that of TorMd method and Win method. After further analysis of the experimental data, it is found that the dependencies obtained from the dependency parsing is likely to be the modification of the context words by the function words such as "一名中医", and the function words "名" depend on the context words "中医". Obviously, when it use HowNet to calculate the similarity of

Table 1 Disambiguation result

	TorMd	Win	Con	CSD	HowCSD
Pmar	0.431	0.449	0.421	0.438	0.471
Increase (%)	4	2.2	5	3.3	0

word sense and concept, the similarity between "名" and "中医" is very small or does not exist, and the word does not play its due indicative role. It can be seen that it is not advisable to filter out function words directly. HowCSD method combines context disambiguation graph and dependency disambiguation graph, which avoids this problem to a certain extent, and its effect is better than TorMd and Win method, with an increase of 4 and 2.2 percentage points respectively. CSD method uses context disambiguation graph and CSD examples better than TorMd but less than HowCSD. This shows that the method of word sense disambiguation based on HowNet and CSD can effectively improve the effect of word sense disambiguation.

4 Conclusion

This paper proposes a word sense disambiguation method based on graph and knowledge base. After dependency parsing of the given ambiguous sentence, the context dependency graph is obtained, and then the context knowledge can be got; use HowNet and the context knowledge to construct the context disambiguation graph; dependency parsing of the examples in HowNet, to get the example dependency graph; get the dependency disambiguation graph by means of the context dependency graph and the example dependency graph; combine the context disambiguation graph and the dependency disambiguation graph, to run graph algorithm to get the right sense for the target ambigous word. The experimental results show that HowCSD can effectively improve the performance of word sense disambiguation.

References

1. Luo, F., Liu, T., Xia, Q., Chang, B., Sui, Z.: Incorporating glosses into neural word sense disambiguation. In: Proceedings of the 56th Annual Meeting of the Association for Computational Linguistics (Volume 1: Long Papers), pp. 2473–2482. Association for Computational Linguistics (2018)
2. Tripodi, R., Navigli, R.: Game theory meets embeddings: A unified framework for word sense disambiguation. In: Proceedings of the 2019 Conference on Empirical Methods in Natural Language Processing and the 9th International Joint Conference on Natural Language Processing (EMNLP-IJCNLP), pp. 88–99. Association for Computational Linguistics (2019)
3. Blevins, T., Zettlemoyer, L.: Moving down the long tail of word sense disambiguation with gloss informed bi-encoders. In: Proceedings of the 58th Annual Meeting of the Association for Computational Linguistics, pp. 1006–1017. Association for Computational Linguistics (2020)

4. Bevilacqua, M., Navigli, R.: Breaking through the 80% glass ceiling: Raising the state of the art in word sense disambiguation by incorporating knowledge graph information. In: Proceedings of the 58th Annual Meeting of the Association for Computational Linguistics, pp. 2854–2864. Association for Computational Linguistics (2020)
5. Li, J.: Research on Chinese Word Sense Disambiguation Method. Tsinghua University, Beijing (1999)
6. Wu, D., Xia, K.: Large-scale automatic extraction of an English-Chinese translation lexicon. Mach. Transl. **9**(3/4), 285–313 (1994)
7. Galley, M., McKeown, K.: Improving word sense disambiguation in lexical chaining. In: Proceedings of the 18th international joint conference on Artificial intelligence, pp. 1486–1488. (2003)
8. Mihalcea, R., Tarau, P., Figa, E.: PageRank on semantic networks, with application to word sense disambiguation. In: COLING 2004: Proceedings of the 20th International Conference on Computational Linguistics, pp. 1126–1132. COLING (2004)
9. Navigli, R., Velardi, P.: Structural semantic interconnections: A knowledge-based approach to word sense disambiguation. IEEE Trans. Pattern Anal. Mach. Intell. **27**(7), 1075–1086 (2005)
10. Agirre, E., Soroa, A.: Personalizing pagerank for word sense disambiguation. In: Proceedings of the 12th Conference of the European Chapter of the Association for Computational Linguistics, pp. 33–41. Association for Computational Linguistics (2009)
11. Agirre, E., de Lacalle, O.L., Soroa, A.: Random walks for knowledge-based word sense disambiguation. Comput. Linguist. **40**(1), 57–84 (2014)
12. Nieto Piña, L., Johansson, R.: Embedding senses for efficient graph-based word sense disambiguation. In: Proceedings of TextGraphs-10: the Workshop on Graph-based Methods for Natural Language Processing, pp. 1–5. Association for Computational Linguistics (2016)
13. Janz, A., Piasecki, M.: Word sense disambiguation based on constrained random walks in linked semantic networks. In: Proceedings of the International Conference on Recent Advances in Natural Language Processing (RANLP 2019), pp. 516–525. INCOMA Ltd. (2019)
14. Scozzafava, F., Maru, M., Brignone, F., Torrisi, G., Navigli, R.: Personalized PageRank with syntagmatic information for multilingual word sense disambiguation. In: Proceedings of the 58th Annual Meeting of the Association for Computational Linguistics: System Demonstrations, pp. 37–46. Association for Computational Linguistics (2020)
15. Yang, Z., Huang, H.: Graph based word sense disambiguation method using distance between words. J. Softw. **23**(4), 776–785 (2012)
16. Lu, W., Huang, H., Wu, H.: Word sense disambiguation with graph model based on domain knowledge. Acta Automatica Sinica. (12), 2836–2850 (2014)
17. Dong, Z.: HowNet. http://www.keenage.com (2013)
18. Hui, W., Weidong, Z., Shiwen, Y., Wu, Y.: The Specification of the Chinese Semantic Dictionary. Peking University, Beijing, China (2011)
19. Tesniâre, L.: Élãments de syntaxe structurale. Klincksieck, Paris (1959)
20. Manning, C.D., Surdeanu, M., Bauer, J., Finkel, J., Bethard, S.J., McClosky, D.: The Stanford CoreNLP natural language processing toolkit. In: Association for Computational Linguistics (ACL) System Demonstrations, pp. 55–60. Association for Computational Linguistics (2014)
21. Liu, Q., Li, S.: Semantic similarity computation based on HowNet. Int. J. Comput. Linguist. Chin. Lang. Process. **7**(2), 59–76 (2002)

Prediction of GDP Carbon Emission Based on Grey Model and Neural Network

Feng Zhang, Huihuang Zhao, and Manimaran Ramasamy

Abstract In order to improve the prediction accuracy of Gross Domestic Product (GDP) carbon emission, combining the advantages of grey prediction and neural network, a discrete grey neural network prediction model named DGMBP is established. This model not only has the advantages of less data sample and larger development coefficient range of GM 1,1 model but also combines the advantages of neural network to deal with nonlinear data sample. Taking the GDP carbon emission data of China, India, the United States and the European Union as an example, the prediction results obtained are stable, and the accuracy of the prediction data is over 95%, which is much higher than that obtained by the linear regression equation and the logical regression equation. Finally, the carbon emission of GDP in 2020–2030 is predicted by using this model.

Keywords GDP carbon emissions · Grey prediction · Neural network

1 Introduction

The theory of grey system was put forward by Professor Deng Julong in 1982. The object of study is a small number of samples with "part information known, part information unknown". The main purpose is to extract effective information from known information and to make scientific predictions about the unknown [1, 2, 3]. Grey prediction model is one of the most active research directions in grey

F. Zhang · H. Zhao (✉)
College of Computer Science and Technology, Hengyang Normal University, Hengyang, China

Hunan Provincial Key Laboratory of Intelligent Information Processing and Application, Hengyang, China
e-mail: happyzhh@hynu.edu.cn

M. Ramasamy
College of Computer Science and Technology, Hengyang Normal University, Hengyang, China

S. Mu et al. (eds.), *4th EAI International Conference on Robotic Sensor Networks*, EAI/Springer Innovations in Communication and Computing,
https://doi.org/10.1007/978-3-030-70451-3_4

system, including the model of GM (1,1), the model of DGM (1,1), the model of DGMC (1, N) and so on [4, 29, 30]. At present, many scholars in many countries and international organizations are studying the grey prediction technology, which has been accepted by more and more people and widely used in agriculture, industry, military and other fields [7, 8, 15, 21, 24, 25]. For example, it is used to forecast the total yield in China's grain development decision-making, to make the best irrigation decision in each place and to add the grey system software to the Management Software Library of IBM as a global service [24, 24]. Use of the grey prediction model has many advantages: data samples larger than or equal to 4 can be modeled and the modeling process is easy to understand [9, 11, 14]. However, the grey prediction model has some defects, and it is not ideal for data with large fluctuation. Therefore, in order to get more accurate results, other models are needed.

The neural network model was first proposed by McCulloch and Pitts in 1943 [13, 18]. Neural networks can process data in parallel and complete nonlinear mapping. It has played an important role in processing inaccurate and incomplete knowledge in the acquisition of information in complex nonlinear systems [19]. In recent years, the research of neural network has attracted the attention of all countries in the world. At present, the application of neural network is more extensive than in the past, and there is a breakthrough in the field of control and decision. For systems with simple data but complex input-output processes, the neural network model works well but requires a sufficiently large data set.

The combination forecasting model of gray neural network emerged in the 1990s. Because the combination forecasting model complemented each other and made up for the shortcomings of the gray forecast and the neural network model, the combination forecasting model has been developed rapidly [26, 27]. So far, some results have been achieved in combination forecasting methods, such as the combination of grey system and regression line moving average model, grey Markoff process, grey Support vector machine model, cluster analysis and neural network [22, 28]. Chen Shuyan and Wang Wei forecast the real-time traffic volume of cross-section motor vehicles on the Beijing-Shijiazhuang expressway by using the combination of the grey system theory and the artificial neural network, the prediction results obtained with a single model are accurate [12].

2 Related Works

With the continuous development of gray prediction models and neural networks, many researchers at home and abroad have begun to explore the combination of other prediction models, and the research models and methods are also different. Sallehuddin R, Shamsuddin SMH, Hashim Szm, et Al. proposed a time series data forecasting method based on grey relational neural network and autoregressive moving average model to forecast national wealth and income under the influence of multi-factors [5]. Karamouz M, Razavi S, Aragejad S used the time lag recurrent neural network method to predict long-term seasonal rainfall [6]. Hsu LC and Wang

CH use grey multivariable model and grey correlation analysis to predict the output of integrated circuits [10]. Hamzacebi C, Akay D, Kubay F compare the direct and iterative neural network methods in multi-period time series [16]. Nai-ming Xie, Si-feng Liu present a discrete grey optimization prediction model [16]. Tsaur, Ruey-Chyn used fuzzy gray regression model to analyze and predict limited time series data. Mehdi Khashei, Seyed Reza Hejazi and Mehdi Bijan proposed a time series forecasting method based on neural network and fuzzy regression model [17]. Mehdi Khashei, Seyed Reza Hejazi, Mehdi Bijan proposed a time series prediction method based on neural networks and fuzzy regression models [20]. Li Bin, Xu Shirong et al. established an optimization model, which used the effectiveness of the forecasting method as an optimization index to solve the weight coefficient of the combined forecasting model [23].

Luo Dang, Shi Yanan, et al. use the results predicted by the GM(1,1) model as the input of the BP neural network, and at the same time the actual value as the output, obtains each hidden layer node's weight value and the threshold value through the training to the data set, then construct the network, and use the prediction result of the GM(1,1) model as the input of the neural network, and perform the simulation to output the predicted value. The gray BP neural network model was established [29]. The grey BP neural network model is established. Wu zhi-zhou and others use the series-type grey neural network forecast model to compare the forecast results under different time intervals and apply it to the point speed forecast of real-time traffic volume, which improves the accuracy of the point speed forecast [30]. He Qingfei and Chen Guiming used the trapezoid formula of numerical integration to construct the background value, improved GM(1,1) and RBF neural network, formed a new series-type grey neural network model and applied it to predict the life of gear pump [31].

Based on the literature, we understand that there are many models for studying carbon emissions, of which the grey prediction model and neural network prediction model are the most widely used. The gray prediction model has high precision and fast calculation ability for data with small calculation amount and small number of samples. The neural network prediction model has strong nonlinear fitting ability and can accurately predict data with irregular changes. Therefore, this paper combines the advantages of the two to ensure the speed of the forecast while ensuring the accuracy of the forecast data. First, we use the GDP carbon emissions data of 2004–2014 to forecast the GDP carbon emissions data of 2015–2018. After analyzing the error between the data and the original data, we determine the accuracy of the model and then predict the GDP carbon emissions of China, India, the United States and the European Union from 2020 to 2030.

In this paper, we propose a model named DGMBP, which is used for data prediction and testing with high accuracy. The main contributions of this paper are summarized as follows:

- A novel model named DGMBP for predicting world's GDP carbon emissions is proposed. We found that few researchers currently focus on the prediction of GDP carbon emissions. The GDP carbon emission is defined as carbon emission per capita/GDP per capita.

- A neural network model is designed for the proposed model solution. Our experimental model combines the advantages of grey prediction and neural network. It not only compares the existing data but also predicts the data. The accuracy of the GDP carbon emission prediction is more than 95%.

It is of great significance to predict the value of GDP carbon emissions. At present, countries in the world are advocating energy conservation and emission reduction. The GDP carbon emission index can fully take into account the different national conditions of the world, so as to distribute the tasks of each country more equitably.

3 Model Construction

3.1 Design of GDP Carbon Dioxide Emissions Prediction Model

The DGMBP is based on a GM (1,1) prediction model which uses the grey prediction; note $s^0(k)$, where $k = 1,2,3, ..., n$. Note $s^{(0)}(1)$ here for the first year and build the following model.

At first, set the known reference data as $s^{(0)} = \{s^{(0)}(1), s^{(0)}(2), s^{(0)}(3), \cdots, s^{(0)}(n)\}$

Add the sequence $s^{(0)}$ once to generate the following vector

$$s^{(1)} = \left\{s^{(1)}(1), s^{(1)}(2), s^{(1)}(3), \cdots, s^{(1)}(n)\right\} \tag{1}$$

$$\text{where,} \quad s^{(1)}(k) = \sum_{i=1}^{n} s^{(0)}(i) \quad k = 1, 2, 3, \cdots, n \tag{2}$$

Generate a sequence by averaging,

$$A^{(1)}(k) : a^{(1)}(k) = \frac{1}{2}s^{(1)}(1) + \frac{1}{2}s^{(1)}(k+1), \quad k = 1, 2, \cdots, n,$$

Then, $A^{(1)} = [a^{(1)}(2), a^{(1)}(3), \ldots, a^{(1)}(n)]$

And create an albino differential equation,

$$\frac{ds^{(1)}}{dt} + zs^{(1)}(k) = y \tag{3}$$

In the formula, z is the development coefficient, and its size reflects the development trend of the sequence, and y is the amount of gray action, which reflects the relationship of data changes.

To find the values of the parameters z and y.

If $v = (z, y)^T$ for reference column,

$$\text{Let } D = \begin{bmatrix} -\frac{1}{2}\left[s^{(1)}(2)+s^{(1)}(3)\right] & , 1 \\ -\frac{1}{2}\left[s^{(1)}(3)+s^{(1)}(4)\right] & , 1 \\ \vdots & \vdots \\ -\frac{1}{2}\left[s^{(1)}(n)+s^{(1)}(n+1)\right] & , 1 \end{bmatrix} = \begin{bmatrix} -a^{(1)}(2), & 1 \\ -a^{(1)}(3), & 1 \\ \vdots & \vdots \\ -a^{(1)}(11), & 1 \end{bmatrix} \quad S_n = \begin{bmatrix} s^{(0)}(0) \\ s^{(0)}(1) \\ \vdots \\ s^{(0)}(n) \end{bmatrix} \tag{4}$$

then the least square method is used to obtain,

$$R\left(\hat{v}\right) = \left(S_n - D\hat{v}\right)^T \left(S_n - D\hat{v}\right) \tag{5}$$

Reaching the minimum value,

$$\hat{v} = \begin{bmatrix} z \\ y \end{bmatrix} = \left(D^{\mathrm{T}}D\right)^{-1} D^{\mathrm{T}} S_n \tag{6}$$

Solving the above differential equation gives,

$$\hat{s}^{(1)}(k+1) = \left(s^{(1)}(1) - \frac{y}{z}\right)e^{-zk} + \frac{y}{z}, \quad k = 1, 2, 3, \cdots, n \tag{7}$$

3.2 Design of Neural Network Prediction Model

Back propagation (BP) neural network is a kind of feedforward neural network, which is made up of a large number of neurons. The strength of the connection is represented by the weight of the neurons in each layer. The BP neural network model designed in this paper is shown in Fig. 1.

The network model includes input nodes, output nodes, and two hidden-layer nodes. Nodes belonging to the same layer have no connection. The input signal passes through each hidden layer node in turn from the input node and is transmitted to the output node. Each output node only affects the output nodes of the next layer. In the network learning process, the learning rules of the steepest descent method are used to continuously modify the weights to obtain the best output.

Back propagation algorithm steps:

According to the description above, we propose a DGMBP algorithm for predicting GDP carbon emissions. The algorithm is as follows:

1. Enter M learning examples $\left(A_k, Z_k^*\right)$, $k = 1,2,3,\ldots,M$.
2. Building a BP network structure.

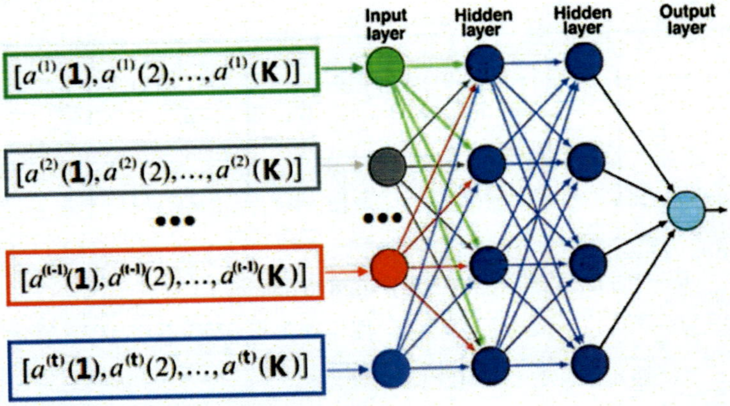

Fig. 1 Design of neural network model

Determine the number of network layers $L \geq 3$ and the number of nodes in each layer. The length of A_k and the nodes of the network input layer are n, and the length q of the learning instance output vector Z_k^* determines that the nodes of the network output layer are q; the number of nodes in the lth layer is $n(l)$. Define the connection weight matrix of each layer, and the connection weight matrix of lth layer to $(l+1)$th layer is $S^{(l)} = \left[S_{ij}^{(l)}\right]_{n^{(l)} \times n^{(l+1)}}$, $l = 1, 2, \ldots, L-1$.

3. Enter the allowable error ε and learning rate η, the number of initial iteration calculations $r = 1$ and the learning instance number $k = 1$.
4. Take the k-th learning example $\left(A_k, Z_k^*\right)$, $A_k = (a_{1k}, a_{2k} \ldots, a_{nk})$, $Z_k^* = \left(z_{1k}^*, z_{2k}^*, \ldots, z_{nk}^*\right)$.
5. The forward propagation calculation is performed by A_k.
   Calculate the output of each node of the input layer: $R_{jk}^l = f\left(x_{jk}\right)$, $(j = 1, 2, \ldots, n)$.
   Calculate the input and output of each layer by layer: $T_{jk}^{(l)} = \sum_{i=1}^{n^{(l-1)}} s_{ij}^{(l-1)} R_{ik}^{(l-1)}$, $R_{ik}^{(l)} = f\left(T_{jk}^{(l)}\right)$, $(l = 1, 2, \ldots, L; j = 1, 2, \ldots, n^{(l)})$.
6. Calculate the error of each output node of the output layer (Lth layer):$z_{jk} = R_{ik}^{(l)}$, $U_{jk} = \frac{1}{2}\left(z_{jk}^* - z_{jk}\right)^2$ $(j = 1, 2, \ldots, m)$.
7. If there is $U_{jk} \leq \varepsilon(j = 1, 2, .., m)$, for any one of the M learning instances k, the learning process ends; otherwise, the error back propagation needs to be performed to modify each connection weight matrix.
8. Error back propagation calculation: modify the connection weight matrix from the hidden layer at the $(L-1)$th layer to the output layer to:

$$\delta_{jk}^{(l)} = -\left(z_{jk}^{*} - z_{jk}\right) f'\left(T_{jk}^{(l)}\right); \quad \Delta s_{ij}^{(l-1)}(t) = \eta \delta_{jk}^{(l+1)} R_{ik}^{(l)}$$

$$s_{ij}^{(l-1)}(t+1) = s_{ij}^{(l-1)}(t) + \Delta s_{ij}^{(l-1)}(t) \quad j = 1, 2, \ldots, m; \; i = 1, 2, \ldots, n^{(l-1)}$$

and modify the connection weight matrix connecting hidden layers layer by layer in reverse:

$$\delta_{jk}^{(l)} = f'\left(R_{jk}^{(l)}\right) \sum_{q=1}^{n^{(l+1)}} \delta_{qk}^{(l+1)} s_{jq}^{(l)}; \quad \Delta s_{ij}^{(l-1)}(t) = \eta \delta_{jk}^{(l+1)} R_{jk}^{(l)}$$

$$s_{ij}^{(l-1)}(t+1) = s_{ij}^{(l-1)}(t) + \Delta s_{ij}^{(l-1)}(t)$$

$$\Big(l = L-1, \ldots, 2, 1; \; j = 1, 2, \ldots, n^{(l)}; \; i = 1, 2, \ldots, n^{(l-1)}$$

9. $k = k + 1(\text{mod}N)$, $t = t + 1$, Go to step (4).

3.3 Prediction Model of DGMBP Based on Combination of GM(1,1) and Neural Network

The realization of DGMBP model is to input training data into GM(1,1) model for fitting prediction and then to take its output as the input of neural network model. A brief flowchart is shown in Fig. 2.

3.4 Testing of DGMBP (Discrete Gray Model with Back Propagation)

In order to measure the performance of DGMBP, the terms of standard deviation ratio and small error frequency are used [12]. The standard deviation ratio is defined as

$$U = \frac{T_2}{T_1} \tag{8}$$

where $T_1 = \sqrt{\frac{1}{n}\sum_{i=1}^{n}\left(s^{(0)}(i) - \overline{s}^{(0)}\right)^2}$, $T_2 = \sqrt{\frac{1}{n}\sum_{i=1}^{n}(e(i) - \overline{e})^2}$, $e(k) = s^{(0)}(k) - \hat{s}^{(0)}(k)$, $\overline{e} = \frac{1}{n}\sum_{i=1}^{n} e(i)$ and $\overline{s}^{(0)} = \frac{1}{n}\sum_{i=1}^{n} s^{(0)}(i)$.

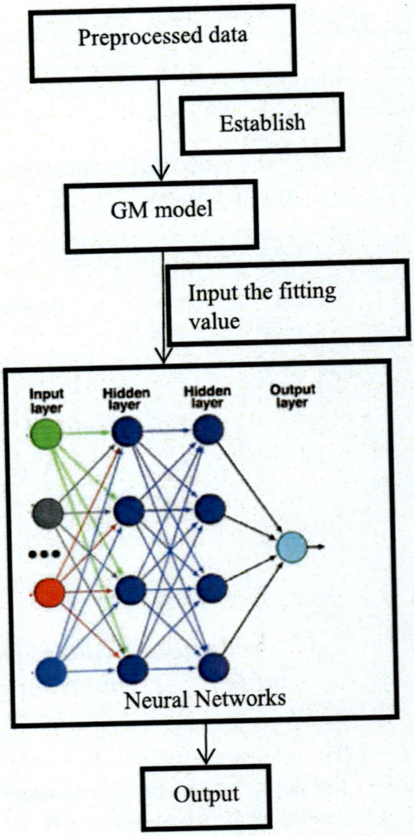

Fig. 2 The flowchart for DGMBP

The small error frequency is defined as

$$R = r\{|e(k) - \overline{e}| < 0.6745T_1\} \tag{9}$$

Based on the accuracy-level division shown in Table 1, the accuracy level of DGMBP is judged. The smaller the value of U is, the smaller the $T1$ is compared with the larger $T1$. The larger the $T1$, the larger the original data dispersion, and the smaller the $T2$, the smaller the dispersion of the prediction error. The larger the value of R, the better. A large value of R indicates that the error is relatively small and the probability of the model is high.

Table 1 Classification of standard of accuracy grade

Model accuracy level	Disqualified (level 1)	Barely qualified (level 2)	Passed (level 3)	Good (level 4)
R	≤0.7	>0.7	>0.8	>0.95
U	≥0.65	<0.65	<0.5	<0.35

Table 2 2004–2014 GDP carbon emissions for four regions (value units * 10^{-4})

Year	2004	2005	2006	2007	2008	2009	2010	2011	2012	2013	2014
China	25.9	24.5	22.9	18.9	15.6	14.9	13.6	12.1	11.1	10	9.1
India	15.9	15.4	13.6	11.7	13	12.7	10.3	9.6	11.1	11	10.8
United States	5	4.8	4.3	4.2	4.1	3.8	3.7	3.6	3.3	3.4	3.3
European Union	3.1	3	2.8	2.4	2.2	2.2	2.3	2.1	2.2	2	1.9

Table 3 Data predicted by DGMBP model (value units*10^{-4})

Country	2015	2016	2017	2018
China	7.3852	6.6576	6.0513	5.6353
India	9.5687	9.4089	9.3180	9.2622
United States	3.0558	2.9656	2.9128	2.8803
European Union	1.8035	1.7619	1.7450	1.7419

4 Experiment

In this paper, the per capita GDP and carbon emissions of China, India, the United States and the European Union from 2004 to 2014 are used as the original data to calculate the GDP carbon emissions, and the GDP carbon emissions from 2015 to 2018 are used as the test data. Data from the World Bank database are shown in Table 2 and the data characteristics show a downward trend, but there are a few outliers, as shown in Table 2.

4.1 Experimental Data

Input the data from 2004 to 2014, use GM(1,1) for data fitting, take the fitting data as the input of the neural network prediction model, forecast the data from 2015 to 2018, and calculate the accuracy of the prediction data. The forecast data for 2015–2018 are shown in Table 3. The overall trend in the image is shown in Fig. 3. The blue part of the curve is the original data and the red part is the prediction of the data.

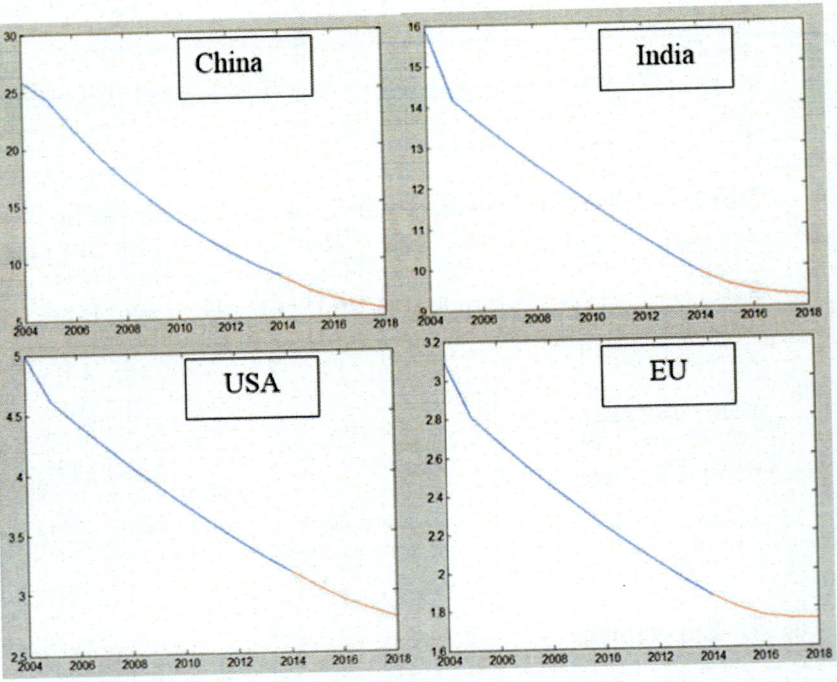

Fig. 3 2004–2018 GDP carbon emission curve for four countries

4.2 Analysis of Experimental Results

In order to verify the validity of DGMBP in predicting regional GDP carbon emissions, a comparative experiment was carried out with linear regression equation and logistic regression equation.

As can be seen from Table 4, by comparing the data accuracy of China, India, the United States and the European Union, it can be found that the DGMBP model has relatively low prediction error and good prediction stability in predicting the data of these four groups, it's over 95% accurate. However, the accuracy of the data predicted by linear regression equation and logical regression equation fluctuates greatly, and the stability of prediction is poor.

4.3 GDP Carbon Emission Forecast

Next, we try to use DGMBP model to predict the trend of GDP carbon emissions around the world. From 2020 to 2030, the GDP carbon emissions of China, India, the United States and the European Union will decrease year by year. As shown in

Table 4 Comparison of accuracy of different methods

Country	Methods											
	Linear regression equation				Logistic regression equation				DGMBP			
	2015	2016	2017	2018	2015	2016	2017	2018	2015	2016	2017	2018
China	0.78	0.59	0.43	0.25	0.66	0.65	0.60	0.55	**0.97**	**0.98**	**0.95**	**0.95**
India	0.82	0.83	0.89	0.77	0.98	0.91	0.80	0.87	**0.99**	**0.99**	**0.99**	**0.96**
United States	0.98	0.99	0.96	0.89	0.81	0.76	0.73	0.67	**0.99**	**0.99**	**0.98**	**0.95**
European Union	0.83	0.78	0.76	0.79	0.95	0.94	0.91	0.82	**0.99**	**0.99**	**0.99**	**0.97**

Amongthe several methods compared, the DGMBP model has the highest prediction accuracy that is the prediction effect of the model is relatively good.

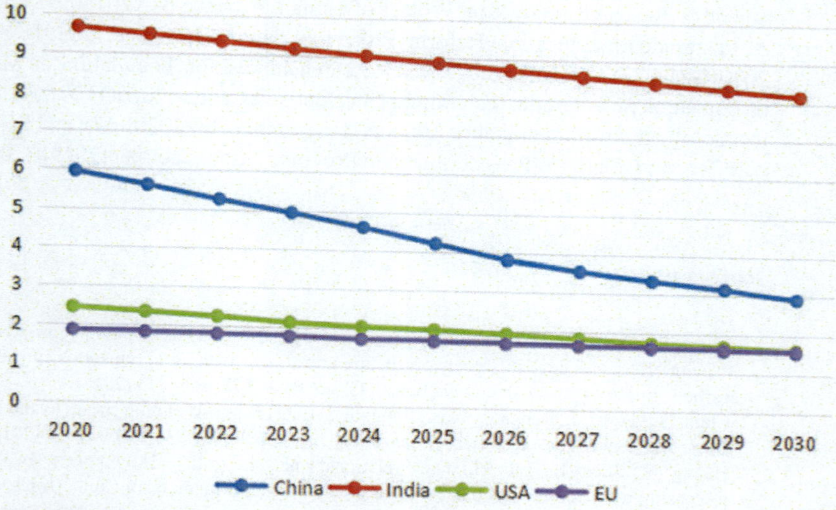

Fig. 4 Prediction of data of 2020–2030 by DGMBP

Fig. 4, Chinese GDP carbon emissions have fallen by the most and the fastest in the past three decades.

5 Conclusions

In this paper, the DGMBP model is used to forecast the carbon emission data of GDP from 2015 to 2018. The forecast accuracy is above 95%, which is higher than that of linear regression equation and logistic regression equation, and closer to the real data. Then, DGMBP model is used to forecast the GDP carbon emission data of China, India, the United States and European Union from 2020 to 2030. The experiment found that China's GDP carbon emissions will decline faster in the next ten years, which is consistent with the Chinese government's promotion of a low-carbon economy. Under the guidance of sustainable development, minimize carbon dioxide emissions and achieve a win-win situation of economic and social development and ecological environmental protection. DGMBP model can provide theoretical support for the realization of low-carbon economy and quantitative basis for the establishment of ecological compensation mechanism.

Funding Statement This work was supported by the National Natural Science Foundation of China (No. 61503128, 61772179), Hunan Provincial Natural Science Foundation of China (No. 2020JJ4152, 2019JJ40005), the Science and Technology Plan Project of Hunan Province (No. 2016TP1020), Double First-Class University Project of Hunan Province (Xiangjiaotong [2018] 469), Postgraduate Scientific Research Innovation Project of Hunan Province (CX20190998), Degree & Postgraduate Education Reform Project of Hunan Province (2019JGYB266, 2020JGZD072), Industry University Research Innovation Foundation of Ministry of Education Science and Technology Development Center (2020QT09), Hengyang technology innovation guidance projects (Hengcaijiaozhi [2020]-67), and Postgraduate Teaching Platform Project of Hunan Province (Xiangjiaotong [2019] 370–321).

References

1. Wu, C.B., Huang, G.H., Xin, B.G., Chen, J.K.: Scenario analysis of carbon emissions' anti-driving effect on Qingdao's energy structure adjustment with an optimization model, part I: carbon emissions peak value prediction. J. Clean. Prod. **172**, 466–474 (2018)
2. Chun, H.E., Qi, L.I., Ranghao, W.U., Bangxin, L.I.U.: Diagnosis of fault circuit by modularized BP neural network based on fault propagation. J. Comput. Appl. **38**(2), 602–609 (2018)
3. Huang, Y., Shen, L., Liu, H.: Grey relational analysis, principal component analysis and forecasting of carbon emissions based on long short-term memory in China. J. Clean. Prod. **209**, 415–423 (2019)
4. Zijian Guo, Mingliang Liu, Bing Li.: Circuit breaker fault analysis based on wavelet packet time-frequency entropy and lm algorithm to optimize BP neural network. The 37th China Control Conference
5. Shuai, S., Yang, L.L., Cao, J.H.: Study on inflfluencing factors of CO_2 emissions from industrial energy consumption: an empirical analysis based on STIRPAT model and industrial sectors' dynamic panel data in Shanghai. J. Finance Econ. **36**(11), 16–27 (2010)
6. Giblin, S., Mcnabola, A.: Modelling the impacts of a carbon emission reduction[J]. Resour. Energy Econ. **32**(3), 310–326 (2009)
7. Hoen, K.M.R., Tan, T., Fransoo, J.C., et al.: Effect of carbon emission regulations on transport mode selection under stochastic demand[J]. Flex. Serv. Manuf. J. **26**(1–2), 170–195 (2014)
8. Schipper, L.: Manufscturing energy use in eight OECD countries: trends through 1988[J]. Energy J. **12**(4), 15–40 (2015)
9. Boyang Li, Chen Liu. Parallel BP neural network on single-chip cloud computer. IEEE International Conference on IEEE International Conference on High Performance Computing & Communications, IEEE International Symposium on Cyberspace Safety & Security. IEEE, 2019
10. Ting, L., Jing-An, S.: Simulation of soil erosion intensity in the three gorges reservoir area using BP neural network. J. Nat. Resour. 04, 131–145 (2018)

11. Wang, J.K., He, Y.L., Lu, H.Y., Li, Z.W.: Study on vibration acceleration prediction model of track inspection vehicle based on BP neural. Network. (2018) https://doi.org/10.1088/1757-899X/435/1/012041
12. Shuyan, C., Wei, W.: Grey neural network forecasting for traffic flow. J. Southeast Univ. Nat. Sci. **034**(004), 541–544 (2004)
13. Ying Shen, ChenMing He, JunHai Cao, Bo Zhang. A relationship construction method between lifecycle cost and indexes of RMSST based on BP neural network. Eng. Asset Manag. 227–233 (2018) https://doi.org/10.1007/978-3-319-62274-3_20
14. Xing-ji Wang. Forecasting construction project cost based on BP neural network. In: International Conference on Measuring Technology & Mechatronics Automation. IEEE Computer Society, 2018
15. Lee, C.T., Hashim, H., Ho, C.S., et al.: Sustaining the low-carbon emission development in Asia and beyond: sustaining energy, water, transportation and low-carbon emission technology[J]. J. Clean. Prod. **146**, 1–13 (2017)
16. Hongliang, X.U.: Prediction model of pipeline corrosion rate based on improved BP neural. Network. **47**(02), 177–181 (2018)
17. Jun, X.I.A.O., Sidong, L.I.U., Liang, H.U., Ying, W.A.N.G.: Filtering method of rock points based on bp neural network and principal component analysis. Front. Comp. Sci. **12**(6), 1–8 (2018)
18. Xie, Z., Zhang, Y., Jin, C.: Prediction of coal spontaneous combustion in Goaf based on the BP neural network. Procedia Eng. **43**, 88–92 (2012)
19. Zhenpeng, Yupeng, Wensheng, Wang, Zhang: DDoS attack detection scheme based on entropy and PSO-BP neural network in SDN. China Commun.: English edition. (7), 144–155 (2019)
20. Wang, B., Pan, X., Li, Y., Sheng, J., Khawaja, F.R.: Road network link prediction model based on subgraph pattern. Int. J. Mod. Phys. **C**(4), 1–27 (2020)
21. Maleki, H., et al.: Air pollution prediction by using an artificial neural network model. Clean Technol. Environ. Policy, 3 (2019)
22. Liang Xie, Jili Tao. Real-time driving pattern prediction based on KPCA and neural network[C]. International Conference on Industrial Artificial Intelligence, 2019.
23. Wang, P., Zhang, X., Han, B., et al.: Prediction model for railway freight volume with GCA-genetic algorithm-generalized neural network: empirical analysis of China[J]. Clust. Comput. **22**(3), 1–10 (2019)
24. Leonardo Felizardo, Afonso Pinto. A Study on Neural Network Architecture Applied to the Prediction of Brazilian Stock Returns. Papers, 2019
25. Xia, Y., Wang, H., Liu, W.: The indirect carbon emission from household consumption in China between 1995–2009 and 2010–2030: a decomposition and prediction analysis. Comput. Ind. Eng. **128**, 264–276 (2019)
26. Xia-mei, Z., Shu-dan, X.I.A., Meng-tao, X.I.E., Ze-liang, Y.U.: Inversion for damping ratio of flflat blade based on BP neural network. Sci. Technol. Eng. **2**(8), 547 (2018)
27. Tao R, Zhang J, Lv Z P, et al. A FCM, Grey model, and BP Neural Network Hybrid Fashion Color Forecasting Method[M]. Knowledge Management in Organizations. 2019
28. Zhao, B., Ren, Y., Gao, D., et al.: Performance ratio prediction of photovoltaic pumping system based on grey clustering and second curvelet neural network[J]. Energy. **171**(Mar. 15), 360–371 (2019)
29. Dang, L., Yanan, S.: Price prediction of commercial housing in Zhengzhou based on grey BP neural network combination model [J]. J. N. China Inst. Water Conserv. Hydrop. **31**(3), 91–92. + 112 (2010)
30. Zhizhou, W., Yujie, F., Wanjing, M.: Point velocity prediction model based on grey neural network [J]. J. Southwest Jiaotong Univ. **47**(02), 285–290 (2012)
31. Qingfei, H., Guiming, C., Xiaohu, C., Chunjiang, Y.: Life prediction of hydraulic pump based on improved grey neural network [J]. China Mech. Eng. **24**(4), 500–506 (2013)

Automatic Optimization of YOLOv3 Based on Particle Swarm Algorithm

Min Hu, Jie Yuan, Xinzhong Zhu, Dongmei Wu, and Gao Hao

Abstract YOLO (look once at a time) is a target detection method of deep learning, which has achieved great success in image classification and localization and has been widely used. YOLOv3 is the third version of YOLO. Compared with previous versions of software, YOLOv3 employs a better basic network (ResNet) and classifier in the network structure. It also uses multi-scale functions for object detection. For a neural network, the setting of hyperparameters is one of the important factors that determine the performance of the model. The hyperparameter of YOLOs is manually designed by experienced researchers. This process requires researchers to constantly turn parameters to enhance network performance and the experimental process is clueless. This chapter presents a method of particle swarm optimization combined with YOLOv3 for automatic parameter tuning. According to the optimization idea of particle swarm algorithm, the optimal hyperparameter is taken as the search target of particle swarm algorithm. Based on a series of comparative experiments, we can prove on the VOC dataset that the YOLOv3 network optimized by PSO achieves higher accuracy than the YOLOv3 network model before optimization.

Keywords PSO · YOLOv3 · Target detection

M. Hu · D. Wu
College of Automation & College of Artificial Intelligence, Nanjing University of Posts and Telecommunications, Nanjing, China
e-mail: wudm@njupt.edu.cn

J. Yuan
Shanghai Aerospace Electronic Technology Institute, Shanghai, China

X. Zhu
Nanjing University of Posts and Telecommunications, Nanjing, China

G. Hao
College of Automation & College of Artificial Intelligence, Nanjing University of Posts and Telecommunications, Nanjing, China

S. Mu et al. (eds.), *4th EAI International Conference on Robotic Sensor Networks*, EAI/Springer Innovations in Communication and Computing,
https://doi.org/10.1007/978-3-030-70451-3_5

1 Introduction

Like other neural networks, the detection performance of YOLOv3 largely depends on the settings of its hyperparameters. In the past, neural networks were tentatively designed by researchers. They tended to pay more attention to the architecture and framework of the network, while ignoring the selection of hyperparameters. In fact, with the development of evolutionary algorithms, neural networks and evolutionary algorithms are getting closer and closer. Many researchers study how to automatically identify appropriate network parameters. Evolutionary algorithms are widely used due to their high robustness and wide applicability. Evolutionary algorithm is the general term for a group of algorithms with different ways of expressing genes [1]. Particle swarm optimization is a typical evolutionary algorithm, and they show a more significant effect on finding the optimal solution of complex problems encountered with traditional calculus and exhaustive methods. Therefore, the field of neural networks and evolutionary algorithms is widely studied, which has been carried out for decades.

With the development of intelligent computing, the evolutionary algorithm combined with the method of optimizing neural network has also made new breakthroughs, which greatly improves the performance of the original neural network, i.e., the chaotic particle swarm algorithm and the differential evolution algorithm. Researchers apply evolutionary algorithms to dynamically select weights during network training or adjust network characteristics to meet specific needs. For example, many scholars combine the reverse propagation algorithm with the evolutionary algorithm. The improved algorithm not only inherits the reverse optimization characteristics of the backpropagation algorithm but also inherits the overall optimization characteristics of the evolutionary algorithm. For example, the particle swarm algorithm searches for the optimal hyperparameters, which significantly improves the performance of the network.

YOLOv3 is an end-to-end object recognition and location algorithm. Its notable feature is that it runs very fast and can reach the standard of real-time detection [2], which has shown its significant advantages in many practical engineering applications and has been practically applied. Object detection tasks include determining the location and category information of objects in the image. However, the accurate realization of the YOLOv3 detection function requires the support of appropriate network hyperparameters, since hyperparameters or network structure parameters show a great influence on the weights obtained by network training. However, the number of network structure parameters and hyperparameters are numerous. It would be inaccurate and time-consuming to rely on the judgment of prior experience [3]. The wide application of target detection and the increasing requirements for accuracy may often increase the burden of researchers. It is obviously difficult for a researcher to constantly waste time trying to find the best YOLOv3 hyperparameters.

Therefore, more and more evolutionary algorithms combine with neural networks to automatically find optimal values of neural structure and hyperparameters.

For example, Sun et al. used Particle Swarm Optimization (PSO) and Bacteria Foraging Optimization (BFO) to solve the problem of insufficient fitting in the target detection Regional Proposal Network (RPN). BFO is used to optimize the parameters in the SVM classification and PSO is used to optimize the loss function. Sun et al. proved that the α and β values in the loss function will affect the sensitivity of the function, and the parameters in the model will automatically find the optimal value by PSO [4]. Compared with the traditional optimization algorithm, the particle swarm optimization algorithm has simple operation and fast convergence speed, and it will be easier to obtain the optimal hyperparameters of the network.

Sun et al. [5] proposed a large-envelope parameter adjustment wavelet neural network control method, in which the particle swarm algorithm is used to realize the adjustment of the large-envelope gain parameter. The simulation results show that the designed wavelet neural network can ensure the control effect and the stability of the system under the condition of untrained equilibrium state with 20% modeling error. And when the aircraft parameters change greatly with the flight status, it still has better control performance. Wang et al. proposed an improved CNN algorithm combining discrete particle swarm optimization and dropout method. To solve the problem of possible gradient explosion and overfitting of the CNN algorithm, the parameters to be trained in PSO algorithm are optimized as particles, and the updated parameters are used for the forward propagation of CNN algorithm, adjusting the network connection weight matrix iteration until the error converges and stops the algorithm. At the same time, comparatively, a dropout layer is added between the output layer and the fully connected layer to prevent co-fitting by making other hidden units unreliable. The recognition rate has been greatly improved with good robustness. Finally, by comparing with the improved WCNN, MLPNN and SVM-ELM algorithm in the data set MNIST and HCL200, the PD-CNN algorithm is higher in recognition rate and convergence speed than the other three algorithms, which proves that the improved algorithm is correct .

This paper applies YOLOv3 algorithm into the image detection and classification problems on the VOC dataset and uses the particle swarm algorithm to find the best hyperparameters suitable for YOLOv3. Hyperparameters such as momentum and learning strategy will affect the accuracy of algorithm detection. We propose and verify this method on the VOC2007 and VOC2012 data sets. This method can automatically search for the best hyperparameters suitable for the YOLOv3 network. This chapter is organized as follows: Sect. 2 briefly reviews YOLOv3; Sect. 3 introduces the particle swarm algorithm; Sect. 4 describes the implementation of the experiment; and Sect. 5 summarizes this research work.

2 A Brief Review on YOLOv3

YOLOv3 is the third-generation version of the YOLO (look only once) series of networks, which is also an end-to-end algorithm used to solve the two problems of location and classification – it solves the regression problem from training to prediction. At the same time, the location and classification information of the target in images can be predicted. In terms of network design, YOLOv3 is different from RCNN, fast RCNN, and faster RCNN. YOLOv3 converts the target detection problem into a loss function regression problem. After inputting the RGB picture to the network, the pre-trained weights can be used to predict the location and category of all objects and the objects' confidence probability. Algorithms such as RCNN divide the problem of classification and positioning into two steps. One is the object category and the other is the resolution of the bounding box. Unlike them, YOLOv3 does not have a process of searching for regional proposals, so its training and prediction are end to end [6]. RCNN and fast RCNN use a selective search method to obtain an initial candidate area, which includes areas where the target may exist in the picture. This selective search method is free from network interference. The difference between faster RCNN and the former algorithm is that it uses RPN (Regional Proposal Network) instead of the former's selective search module. The RPN network uses softmax to determine the fit between the obtained anchor and the actual target, so as to analyze whether RCNN or faster RCNN requires two steps to complete target detection. This is one of the biggest differences between the above networks and the YOLOv3 network. YOLOv3 draws on the GoogleNet classification network structure. Instead of the inception module, it uses the convolutional layers of size 1 * 1 and 3 * 3 respectively. The existence of the convolutional layer is to unify cross-channel information. In addition, modified on the basis of the loss function of YOLOv2, the biggest change in YOLOv3 is to replace the classification loss with bipartite cross entropy, that is, the network output S * S * (B * 5 + C) size vector. S, B, and C represent grid size, box size, and the number of types, respectively.

The YOLOv3 network structure (Fig. 1) uses a full convolutional layer. In the figure, DBL is the smallest component of the network structure, Res is the residual network module, and Concat is the channel-dependent module. On the basis of Darknet-53, YOLOv3 outputs three feature maps of different sizes through up-sampling and Concat operation, corresponding to deep, medium, and shallow features from top to bottom, realizing the fusion of deep and shallow features [7].

3 The Proposed Particle Swarm Optimization

3.1 Overview

Particle swarm algorithm is a biological heuristic method in the field of computational intelligence, and it is a kind of evolutionary algorithm. The idea of the algorithm originates from the predation behavior of birds. When birds are preying, the simplest strategy to find food is to search for the bird closest to the food. The abstract analogy of birds to particles is the same. The particle swarm algorithm uses the disorder in the particle space to continuously update the position of the

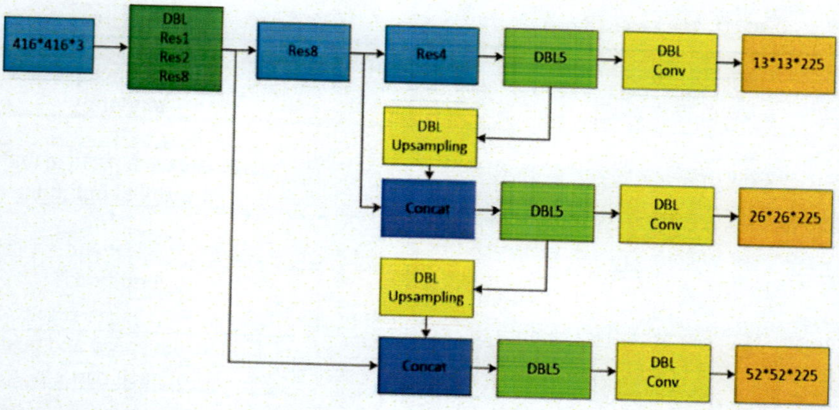

Fig. 1 YOLOv3 network structure

particles, forcing the particles to constantly approach the optimal value, so as to obtain the optimal parameters with the highest fitness function. Give a set of random solutions at the beginning of the algorithm and then search for the best value through an iterative method [8]. Compared with other evolutionary algorithms, particle swarm algorithm is easier to implement. In reality, a large number of problems can be converted into function optimization problems, so optimization algorithms are widely used. The particle swarm algorithm can achieve efficient optimization of high-dimensional complex functions [9]. The fitness function of the particle swarm algorithm is determined by the optimization function, and the speed of each particle determines its update direction. The general flowchart of the particle swarm algorithm is shown in Fig. 2.

In this paper, a training experiment of YOLOv3 is performed on the VOC data set, and the loss function value of the network under the same number of iterations is used as the individual's fitness score. After multiple rounds of iteration, partial optimal values (pbest) and global optimal values (gbest) are obtained and their corresponding hyperparameters are saved [10].

3.2 Hyperparameter Initialization

The particle swarm algorithm first randomly sets a group of hyperparameters within the limited range as shown in Table 1 and feeds them to the network for training to obtain the loss value [11] – the loss value is the fitness function value – and the parameter corresponding to the best fitness value is recorded. In each iteration, the particles keep optimizing in the direction of these two parameters, and the new parameters obtained are fed to the network in turn and iteratively keeps updating these two parameters until the end of the iteration. Figure 3 shows the

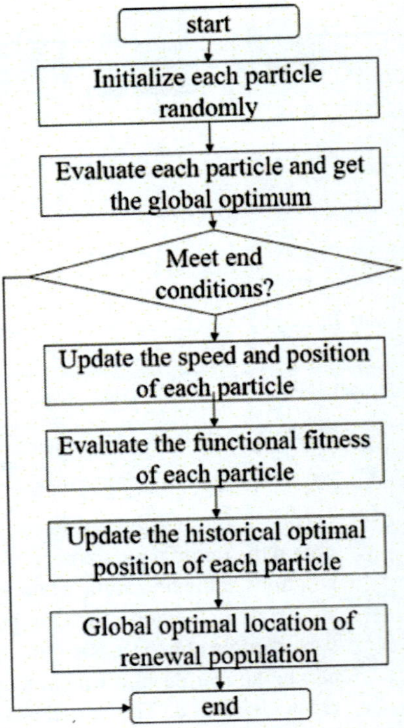

Fig. 2 The general process of particle swarm optimization

Fig. 3 Initial hyperparameters of the YOLOv3 network

LR	MO	BI	ST1	ST2
1e-3	0. 9	5e+2	4e+2	7e+2

Table 1 The range of different hyperparameter values with particle swarm optimization

Hyperparameter	Values
LR	0.001–0.1
MO	0.85–0.95
BI	400–700
ST1	300–499
ST2	500–800

hyperparameters of YOLOv3 and the corresponding original parameters that need to be optimized in this paper.

In Fig. 3, LR represents learning rate, MO represents momentum coefficient, and BI represents "burn in." When the number of iterations is less than "burn in," the learning rate is updated in one way and when it is greater than "burn in," the policy update method is used. ST1 and ST2 are Step1 and Step2, respectively. When iterating to Step1 and Step2, the learning rate will be reduced to one-tenth respectively. According to the past experience of neural network image classification, the range of hyperparameters is limited (see Table 1). The purpose of this is to prevent the particle swarm algorithm from being too large when searching for updated parameters and causing the loss function to fail to fit. The particle swarm algorithm first generates random numbers in the range to obtain the initial parameters and then iteratively updates the parameters according to some rules (described below) to create the YOLOV3 architecture [12].

The overall structure of YOLOv3 is composed of alternating combinations of convolutional layers and pooling layers [13]. In addition, when initializing particles, the number of particles (total number of search particles) and the particle search dimension (the number of hyperparameters to be optimized) must be determined in advance. Then, give each particle an initial location and velocity within the range. The rules for creating YOLOV3 architecture for selected hyperparameters are as follows:

- When the YOLOv3 model is first trained, the initial value of LR is 0.001. Generally, the range of LR is between 0.001 and 0.1. As the number of iterations increases, the learning rate is reduced to accelerate the convergence of the loss function. Throughout the training period, the learning rate from the start to the end of the training should be reduced by one hundred times the initial value.
- Batch represents the number of samples of a batch of training samples [14]. The parameters are updated for every batch of samples to ensure training speed and computing power. Generally, 8 or 64 is selected.
- With the training iteration, the loss function will converge slowly in the iterative process at the later stage of the training. Momentum can better solve this problem [15]. The range of momentum is generally 0.85 to 0.95, and the momentum cannot be greater than 1.
- When the number of iterations reaches burn_in, the update method of the learning rate will be changed. The initial network value is 500, and the range is limited to 400 to 700.
- When iterating to Step1 and Step, the learning rate will become one-tenth of the original, which is initially set to 400 and 700, respectively, and the range is limited.

3.3 *Brief Description of Algorithm Formula*

The particle swarm algorithm first generates initial values through random numbers and obtains the corresponding fitness function values through the initial values. With continuous iteration, partial optimal value and global optimal value are obtained by comparing the size of fitness function. The partial optimal value is referred to as pbest and the global optimal value is referred to as gbest. The particles will search for the optimal value along the direction of these two optimal values and iteratively update according to the following formula [16]:

$$s_i = w * s_i + c1 * \text{rand}\,() * \left(\text{pbest}_i\text{-}l_i\right) + c2 * \text{rand}\,() * \left(\text{gbest}_i - l_i\right) \quad (1)$$

$$l_i = l_i + s_i \quad (2)$$

Where, l_i is the current hyperparameter, s_i is the current hyperparameter adjustment direction, w is the inertia weight, and gbest and pbest are the global optimal number and the local optimal number, respectively. rand() is a random number between 0 and 1. $c1$, $c2$ are learning factors, usually $c1 = c2 = 2$ [17].

3.4 Fitness Function

The loss function of YOLOv3 is taken as the fitness function of the particles. Because a lower loss function value is required, the fitness score is in reverse order. The hyperparameter corresponding to the smallest loss function value is the global optimal parameter.

3.5 Particle Tuning Iteration

With continuous iteration, the fitness function value of all particles is obtained in each iteration, and the fitness value of each particle is compared with the previous particle in turn and the smaller value is selected. Then, find an optimal individual from each iteration of the particle swarm and an optimal individual in all iterations. Put the above two particles into the particle update formula of the next iteration, that is, search for more excellent parameters along their directions [18].

Randomly select a hyperparameter and update the speed direction from the range of the selected hyperparameter. In each iteration, the fitness value of each particle is compared with the current optimal parameters [19]. If this particle is better, use it as the current best position (pbest) and update gbest in the same way. The iteration termination condition selects a value according to the range of the set hyperparameter and is usually selected as the maximum number of iterations so far or the best position that is searched by the particle swarm so far to meet the predetermined minimum fitness function value.

4 Experiment Implementation

4.1 Dataset

The experiment in this article uses the official data set of VOC2007. VOC contains two types of label data: detection and semantics. All of the data have detected labels and some also include semantic labels. VOC datasets are currently widely used to judge the performance of a network algorithm. Many researchers have reported their performance on these data sets. The data set objects include 20 categories, including 9963 labeled images, which consist of three parts – train, validation, test – and 24,640 objects are labeled [20].

4.2 Experimental Facility

In this experiment, all training on YOLOv3 uses the network architecture of the YOLO official website. We use Nvidia Geforce 2080TI to run our experiments to adjust the hyperparameters of YOLOv3 and always maintain the unity of the hardware.

4.3 Obtain Optimal Hyperparameters

This paper mainly introduces a method of iteratively obtaining the best hyperparameter combination of YOLOv3 network structure using particle swarm optimization algorithm, so as to obtain a lower loss function value and accuracy [21–24].

The initialization parameters of our particle swarm algorithm are set as follows: the maximum iterations is 10, the number of initial particles is 10, the search dimension dim = 5, initial inertia weight $w = 0.8$, $c1 = 2$, and $c2 = 2$.

The experiment uses particle swarm optimization to optimize some YOLOv3 hyperparameters, while other hyperparameters (such as normalization) remain unchanged. In fact, in order to ensure that in a training process. These hyperparameters to be optimized need to be iterated within a limited range.

In this experiment, the number of particle swarms is 10, the number of iterations is 10, the parameters to be optimized are 5, and the loss function of YOLOv3 is selected as the fitness function. After each hyperparameter adjustment, feed the hyperparameters into the YOLOv3 network for a complete training. The initial number of iterations of YOLOv3 is 520,000. Since it is found in actual experiments that the model has basically converged after 1000 iterations, very large training iterations will only consume computing resources, so the training period of this experiment is set to 1000 iterations.

Table 2 Part optimal hyperparameters of particle swarm optimization algorithm after 10 iterations

LR	MO	BI	ST1	ST2	Loss
0.007	0.92	424	400	800	0.144346
0.007	0.9	443	363	700	0.159003
0.007	0.9	460	350	740	0.13371
0.005	0.89	304	300	800	0.157156
0.008	0.88	448	347	800	0.144388
0.005	0.9	461	367	800	0.150431
0.005	0.89	453	363	700	0.156863
0.007	0.89	408	474	700	0.143862
0.004	0.93	451	300	700	0.155275
0.008	0.9	391	442	800	0.14697

4.4 Experiment Results

We propose the particle swarm algorithm to optimize YOLOv3 hyperparameters. This is a method that is different from the traditional method. The main purpose is to automatically search for the optimal hyperparameters suitable for the network and ultimately improve the detection performance of the network. The experiment verifies the effect of the experiment through the general VOC data set.

In this experiment, the number of particles is 10. First, randomly generate the location of 10 particles within the range as shown in Table 1 and put the initial hyperparameters of the 10 particles into YOLOv3 training to obtain a set of initial local optimal values. Then, after the first iteration, according to the particle swarm formula (see formulas (1), (2)), a set of new hyperparameters of 10 particles are obtained and they are put into YOLOv3 for training again. If there is a better fitness value, replace the parameter at the corresponding position. With this reciprocation, after 10 iterations, a set of the latest local optimal values are obtained (see Table 2). With the iteration of particles, the particle with the best fitness function value is called the global best particle, that is, the best hyperparameter at the end of the experiment.

In order to demonstrate the advantages of combining particle swarm optimization and YOLOv3 network integration and prevent data distortion caused by errors, under the same experimental environment, the original hyperparameters and the global optimal hyperparameters were fed to YOLOv3 five times respectively. Comparing the five data with the network prediction after tuning, experimental results (see Fig. 4) show that the global optimal hyperparameter has a smaller loss function value. The calculated average loss function of the original data is 0.31554 and the average loss function after optimization is 0.1373. The prediction accuracy has been greatly improved.

In the entire training, the experiment requires a total of 100 YOLOv3 1000 iterations of training. In the same experimental environment, it can be seen from Fig. 3 that the YOLOv3 network combined with the particle swarm algorithm uses more excellent hyperparameters to greatly improve the convergence speed of the loss function. The function value is smaller under the same number of iterations.

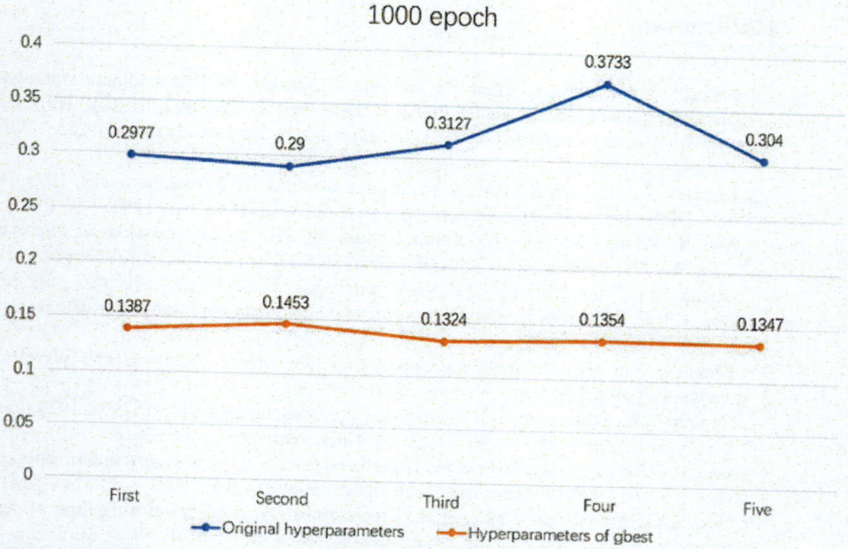

Fig. 4 The performance of original parameters and global optimal parameters in YOLOv3

5 Conclusion

We propose a deep evolution method to automatically discover the best hyperparameter combination of the YOLOv3 model. Use the excellent performance of particle swarm algorithm to find the most suitable hyperparameters to indirectly improve the performance of the YOLOv3 network. It not only decreases the labor cost of the researcher to manually adjust the hyperparameters, but it is also more scientific than manual adjustment of the hyperparameters rather than empirical. In this paper, a total of five hyperparameters in the YOLOv3 network are adjusted. The results of the loss function after training show the loss function value is greatly reduced and the detection accuracy is also increased. The experimental results, compared with the original hyperparameters, show that use of the particle swarm algorithm to search for the best hyperparameters suitable for the network can effectively reduce the value of the YOLOv3 loss function, thereby improving the detection performance of the network.

Acknowledgments The authors acknowledge the support by the National Nature Science Foundation of China (No. 61931012) and Open Program of National Key Laboratory of Science and Technology on Space Intelligent Control (No. 6142208190105, 6142208180, and KGJZDSYS-2018-02).

References

1. Bakhshi, A., Noman, N., Chen, Z., Zamani, M., Chalup, S.: Fast automatic optimisation of CNN architectures for image classification using genetic algorithm. In: 2019 IEEE Congress on Evolutionary Computation (CEC), Wellington, New Zealand, 2019, pp. 1283–1290, https://doi.org/10.1109/CEC.2019.8790197
2. Kennedy, J., Eberhart, R.: Particle swarm optimization. In: Neural Networks, 1995. Proceedings, IEEE International Conference on, vol. 4, pp. 1942–1948, Nov 1995.
3. Xie, Y., Zhang, W.: A novel tuning method for PID controller based on improved PSO algorithm for unstable plants with time delay. In: 2019 Chinese Control Conference (CCC), Guangzhou, China, 2019, pp. 4270–4275, doi: https://doi.org/10.23919/ChiCC.2019.8865547.
4. Box, G.E.P., Wilson, K.B.: On the experimental attainment optimum conditions [J]. J. R. Stat. Soc. B. **13**(1), 1–45 (1951)
5. Bingbing Xu, Huawei Shen, Qi Cao, Yunqi Qiu, Xueqi Cheng. Graph Wavelet Neural Network'2019 ICLR
6. Campbell, R., Wilkinson, J.: Hydrocarbon gas processing [P]. US: 4157904, 1979-06-12
7. LNCS Homepage. http://www.springer.com/lncs. Accessed 21 Nov 2016.
8. Cao, B., Zhao, J.W., Lu, Z.H., et al.: Distributed parallel particle swarm optimization for multi objective and many objective large scale optimization [J]. IEEE Access. **5**, 8214–8221 (2017)
9. Hou, C.G., Yu, X., Cao, Y.D., et al.: Prediction of synchronous closing time of permanent magnetic actuator for vacuum circuit breaker based on PSO-BP [J]. IEEE Trans. Dielectr. Electr. Insulat. **24**(6), 3321–3326 (2017)
10. Everingham, M., Van Gool, L., Williams, C.K., Winn, J., Zisserman, A.: The Pascal visual object classes (voc) challenge. Int. J. Comput. Vis. **88**(2), 303–338 (2010). 6
11. Cui, S.J., Li, J.: The identification of mature apple based on the chromatic aberration[J]. J. Northwest Univ. (Nat. Sci. Ed.). **46**(6), 993–997 (2011) (in Chinese)
12. Li, C., Song, D., Tong, R., et al.: Illumination-aware faster R-CNN for robust multispectral pedestrian detection [J]. Pattern Recogn. **85**, 161–171 (2019)
13. Bodla, N., Singh, B., Chellappa, R., et al.: Soft-NMS – improving object detection with one line of code[C]. In: IEEE International Conference on Computer Vision, IEEE, 2017, pp. 472–480
14. X. Gibert, V.M. Patel, R. Chellappa. Deep multitask learning for railway track inspection[J]. IEEE Trans. Intell. Transport. Syst., 2016:1–12
15. Krasin, I., Duerig, T., Alldrin, N., Ferrari, V., Abu-El-Haija, S., Kuznetsova, A., Rom, H., Uijlings, J., Popov, S., Veit, A., Belongie, S., Gomes, V., Gupta, A., Sun, C., Chechik, G., Cai, D., Feng, Z., Narayanan, D., Murphy, K.: Openimages: A public dataset for large-scale multi-label and multi-class image classification. Dataset available from https://github.com/openimages, 2017. 2
16. Shrivastava, A., Sukthankar, R., Malik, J., Gupta, A.: Beyond skip connections: Top-down modulation for object detection. arXiv preprint arXiv:1612.06851, 2016. 3
17. Lin, T.-Y., Maire, M., Belongie, S., Hays, J., Perona, P., Ramanan, D., Dollár, P., Zitnick, C.L.: Microsoft coco: Common objects in context. In: European conference on computer vision, pp. 740–755. Springer, 2014. 2
18. He, K., Zhang, X., Ren, S., Sun, J.: Deep residual learning for image recognition. In: Proceedings of the IEEE Conference on Computer Vision and Pattern Recognition, pp. 770–778, 2016. 3
19. Zhao, L., Li, S.: Object detection algorithm based on improved YOLOv3. Electronics. **9**, 537 (2020)
20. Redmonand, J., Farhadi, A.: Yolo9000: Better, faster, stronger. In: Computer Vision and Pattern Recognition (CVPR), 2017 IEEE Conference on, pp. 6517–6525. IEEE, 2017. 1, 2, 3
21. Szegedy, C., Ioffe, S., Vanhoucke, V., Alemi, A.A.: Inception-v4, inception-resnet and the impact of residual connections on learning. 2017. 3

22. Lu, H.M., Zhang, M., Xu, X., et al.: Deep fuzzy hashing network for efficient image retrieval. IEEE Tran. Fuzzy Syst. (2020). https://doi.org/10.1109/TFUZZ.2020.2984991
23. Lu, H.M., Li, Y.J., Mu, S.L., et al.: Motor anomaly detection for unmanned aerial vehicles using reinforcement learning. IEEE Internet Things J. **5**(4), 2315–2322 (2017)
24. Lu, H.M., Zhang, Y., Li, Y.J., et al.: User-oriented virtual mobile network resource management for vehicle communications. IEEE Trans. Intell. Transport. Syst. (2020). https://doi.org/10.1109/TITS.2020.2991766

Contour Mask and Matting Driven Face Image Generation

Xin Jin, Zhonglan Li, Xiaodong Li, Mingxue Yu, and Quan Zhou

Abstract This chapter proposes a face generation model based on contour mask and matting, which can well control the distribution quality of generated images. In the generated face model, a mask labeled with a face contour is input to ensure that the generated face conforms to the distribution of the corresponding face contour, and the distribution quality of the generated image is improved. Face generation based on contour mask mainly considers the generation of contour information provided by the mask and the generation of style information composed of other parts of the face. In this chapter, masks are considered as two types of semantic labels with face and background images. This guarantees the accuracy of semantic contour features, and then performs detailed style processing on faces to generate high-quality pictures. Therefore, the generator has both semantic generation and meticulous style information. Finally, inspired by image matting, this chapter improves the face generation model based on contour mask, adds matting information to the input and constraints, and improves the generation effect of the edge part of the face. By comparing with existing unconditional face generation models and semantic label-guided face generation models on the same dataset, it is found that our model proposed in this chapter performs better.

Keywords Face generation · Contour mask · Matting · Generative adversarial networks

X. Jin · Z. Li · X. Li (✉)
Beijing Electronic Science and Technology Institute, Beijing, P.R. China

M. Yu
Xidian University, Xi'an, P.R. China

Q. Zhou
Nanjing University of Posts and Telecommunications, Nanjing, P.R. China

S. Mu et al. (eds.), *4th EAI International Conference on Robotic Sensor Networks*, EAI/Springer Innovations in Communication and Computing,
https://doi.org/10.1007/978-3-030-70451-3_6

1 Introduction

Traditional artificial intelligence technology has a wide range of applications in virtual network resource management [1], anomaly detection for unmanned aerial vehicles (UAVs) [2], and efficient image retrieval [3]. But in the field of image generation, face image generation is a big challenge, and there are very wide application scenarios. In recent years, research on related technologies of human faces has been very hot. However, the lack of face datasets has also greatly restricted the development of related deep learning technologies such as face detection, and current dataset collection methods are difficult to guarantee the quality and distribution of datasets. First, using GAN to generate faces can solve these problems and face image generation can ensure that the number of datasets is large enough, the generation of noise or mask is very simple and can be generated in batches. Second, the distribution of the dataset can be guaranteed, because the generation of noise and mask can be controlled.

Face image generation is different from street scene generation. It has higher complexity, and face image generation requires stricter details. In this chapter, we try to generate a face from a contour mask using a conditional GAN with large constraint. To generate a face from a contour mask, first, the contour information of the face is given to the network. Compared with the semantic label, the semantic and location information of the detailed part is reduced. Secondly, with strong constraints, the discriminator needs to determine the correspondence between the mask map and the generated picture, that is, the generator needs to learn the corresponding representation for each face shape (contour map). Inspired by image matting, this chapter improves the face image generation model based on contour mask, adds matting information to the input and constraints. On the premise of ensuring that the generated face is realistic, the details such as hair have a better generation effect.

The contributions of our method are as follows:

1. This is the first work that can generate a face image using only a facial contour without any facial parts.
2. This chapter proposes a face image generation model based on contour mask, which can well control the distribution quality of the generated images. In the generated face model, a mask labeled with a face contour is input to ensure that the generated face conforms to the distribution of the corresponding face contour, which improves the distribution quality of the generated image.
3. Inspired by image matting, the face generation model based on contour mask is improved. Matting information is added to the input and constraints. Under the premise of ensuring that the generated face image is realistic, in the details such as hair has a better generation effect.

2 Related Work

2.1 DCGAN

DCGAN [4] Previously, convolutional neural networks were used for classification tasks, which were difficult to apply to generative models and difficult to apply to GANs. DCGAN [4] determined the basic architecture of the convolutional GAN, which is capable of stable training in most settings.

2.2 PGGAN

There are two problems with high-resolution image generation. One is that the high-resolution image during training will enlarge the difference between the training distribution and the generated distribution, thereby magnifying the gradient problem. Second, due to the limitation of the existing memory, it is difficult to truly implement large batch training, which causes the distribution of one batch to be difficult to represent the distribution of the entire training set when the gradient of a small batch is reduced, resulting in unstable training. In response to the above problems, PGGAN [5] used progressive training for the first time to achieve the transition from low resolution to high resolution. The progressive training method can improve the stability and speed of training, thereby improving the quality of training. For the first time, PGGAN [5] realized face image generation from noise.

2.3 StyleGAN

The traditional unconditional GAN image generation is essentially a parameter space composed of network layer parameters to fit the distribution of the dataset. It is impossible to control the generated image. StyleGAN [6] borrowed the idea of style transfer and regarded image generation as constantly adding style to images. At the same time, in order to better control the image generation, StyleGAN [6] generated a style space to better decouple the style. The appropriate generation of the hidden vector can be used to perform style fusion to control the image generation.

PGGAN [5] and StyleGAN [6] both solve the unconditional GAN problem, and both start from a hidden vector to generate a high-definition human face. Constraints generated by unconditional GANs are small. During training, the discriminator only needs to determine whether the generated image meets the true distribution, but since the real distribution is broad (real faces that humans can perceive), the generator network parameters can more easily fit the real distribution. However, unconditional GAN is difficult to control the appearance of face generation (although StyleGAN [6] can combine different faces through style fusion), and cannot guarantee the quality of image generation. In order to better control the distribution quality of the generated image, this chapter proposes a face image generation model based on contour mask and matting. In the model, a mask labeled with the contour of the face is input. This ensures that the generated face conforms to the distribution of the corresponding face contour so that the distribution quality of generated image is improved.

3 Face Generation Networks

3.1 Architecture

Face image generation based on contour mask divides the face style information into two types, one is the contour information provided by the mask, and the other is the style information composed of other parts of the face. We consider generating two types of information separately.

1. Contour information is equivalent to rough style. Inspired by conditional GAN generation with semantic label graphs, we consider masks as two types of semantic label graphs, one is a face image and the other is a background image. Similar to the pix2pix [7] model, we use mask maps labeled with two types of semantics as input to the generator, downsampling through convolution, expanding the dimensions, and then continuously convolving to expand the information, which ensures the semantics accuracy of contour features. In addition, pix2pix [7] can learn some rough and medium styles with large parameters through convolution.
2. Face generation requires higher precision and fidelity, and the requirements for detail style are more stringent. In the streetscape generation of pix2pix [7], because there are many and simple categories, the details within the categories do not change significantly, and they can be successfully generated. We need to focus on the generation of detailed information (meticulous style) based on pix2pix because face generation only focuses on one face category and there are changes in details such as eyes, nose, eyebrows, and wrinkles. Refer to StyleGAN [6] for detailed style added to the scale, and pix2pix [7] is three times or four times downsampling of the picture, which is exactly on the scale interval of the detailed style, so it can be convolved after downsampling. In the process, we can add meticulous style information like StyleGAN [6].

Based on the above analysis, this chapter designs a face generation model based on contour mask. The overall design diagram is shown in Fig. 1.

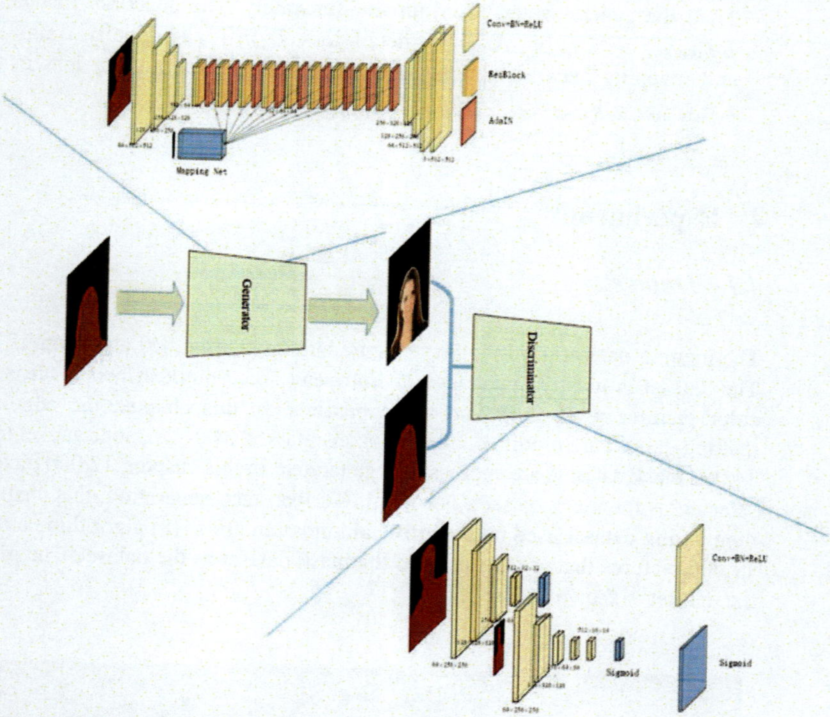

Fig. 1 Overall design drawing of face image generation model based on contour mask

3.2 *Training*

The specific training process is as follows:

1. Mask as a semantic label map is first sent to the discriminator with the fake picture generated by the generator, and the discriminator predicts the fake picture as a false loss $L_{fakefake}$.
2. The mask is sent to the discriminator together with the real picture as a semantic label map, and the loss $L_{truetrue}$ caused by the discriminator predicting the real picture as true is obtained, and the result of the four-layer downsampling convolution is saved.
3. The semantic segmentation map and the fake picture are sent to the discriminator again to obtain the loss $L_{faketrue}$ caused by the discriminator predicting the fake picture as true, and the result of the four-layer downsampling convolution is saved.
4. Compare the two saved results with L1loss to calculate the detail loss L_{detail}.

5. Train the generator and discriminator separately. The generator loss is: $L_g = L_{faketrue} + L_{detail}$. And the detail loss L_{detail} additionally trains AdaIN and mapping network parameters once, and the discriminator loss is: $L_d = L_{fakefake} + L_{truetrue}$.

4 Experiments

4.1 Datasets

The matting dataset used by the semantic human matting [8] algorithm is 34,000. The dataset is mostly upper face of European and American people. It is mainly color pictures with black and white pictures. In this chapter, the edges of the matting dataset are modified, and the pixels marked as 0–1 regions are set to 1, and 34,000 mask datasets are successfully generated. In this chapter, 12,000 pictures are selected as the mask dataset. For matting-driven face generation, this chapter uses the matting dataset used by the artificial human matting [8] algorithm, and selects 11,000 pictures that are the same as the mask dataset as the dataset. The picture of the dataset is shown in Fig. 2.

Fig. 2 Mask dataset display

4.2 Implementation Details of Face Image Generation Based on Contour Mask

In this chapter, LSGAN [9] is used for stable training. The specific experimental parameters are a training batch of 8, initial learning rate of 0.0002, a total of 200 rounds of training, each round of learning rate drops by 0.1, and the optimization algorithm used is Adam [10] Optimization algorithm, whose momentum parameter is 0.5. In the loss function, 10 patchGAN losses are used, the perceptual loss uses the contrast loss L1 of different layers of VGG [11], and VGG [11] uses a pre-trained network. The discriminator loss composed of two parts accounts for 0.5, and the generator loss composed of three parts accounts for 1/3.

4.3 Implementation Details of Face Image Generation Based on Matting

It is the same as the mask-generated face generation model, but the difference is the face generation loss function under matting guidance. The 6 patchGAN loss is used, and the perceptual loss is the contrast loss L1, VGG [11] of different layers. VGG [11] uses a pre-trained network. The discriminator loss remains the same during the training process and consists of two parts, each with a ratio of 0.5. The generator loss is divided into three stages during the training process. During the training process before, during, and after the training, the coefficients of the overall loss are 0.8, 0.5, and 0.4, while the coefficients of the loss of detail are 0.2, 0.5, and 0.6.

4.4 Experimental Results

In this chapter, the results of the face generation model based on contour mask proposed are shown in Fig. 3, where the upper one is the input and the lower one is the generated face.

As can be seen from Fig. 3, the mask-based face generation model proposed in this chapter has a good generation effect. The generated face model has symmetrical features of the face, such as eyes, nose, mouth, and other face parts generate a good effect. And the details are very natural, achieving the expected goal.

We analyze and compare the trend chart of the loss function during training, as shown in Fig. 4. Compared with the loss chart of the previous model, the model generator and discriminator confrontation process in this chapter is more ideal, and the generator loss is closer to the discriminator loss, which indicates that the real image is closer to the generated image, and at the same time, the discriminator loss

Fig. 3 Face generation model effect based on contour mask

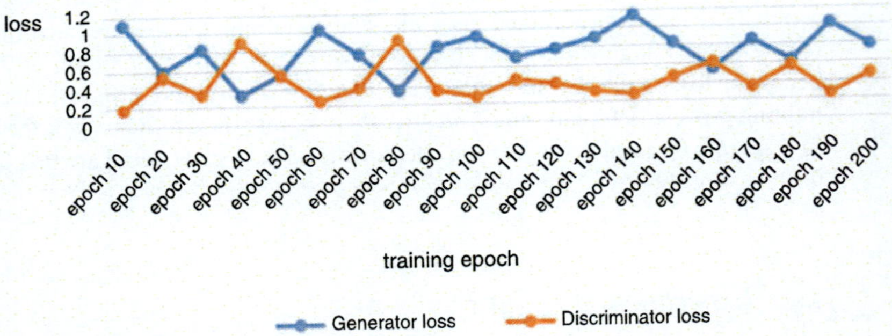

Fig. 4 Facial training model training loss chart based on contour mask

floating on 0.5, the generation effect is ideal, this is because the model adds details as a style, which makes the details clearer.

For face image generation based on matting, Fig. 5 shows the experimental results (the above is the face under the guidance of matting, the following is face image generation guided by contour mask). As can be seen from Fig. 5, face image generation based on matting guarantees that the generated images are realistic, and has better generation effects in details such as hair while in mask-guided face generation, hairline edge generation is stiff.

Fig. 5 Face generation model generation effect based on matting

5 Conclusion and Discussion

Based on unconditional face generation and face generation guided by semantic labels, this chapter creatively uses masks for conditional face generation. In the high-definition face generation model based on the mask, this chapter injects nose, hair, eyes, etc. as detailed styles. How to further distinguish the thickness and granularity of these styles and explore where to inject are the next steps. For the mask image with natural blurring defects at the edges, this chapter improves the mask high-definition face generation model and designs a matting-guided high-resolution face generation model. Finally, it is proved through experiments that the generation effect is clearer and better in indicators than other generation models.

Acknowledgments This work is partially supported by the National Natural Science Foundation of China (grant numbers 62072014, 61701008), the Beijing Natural Science Foundation (grant number L192040), the Open Project Program of State Key Laboratory of Virtual Reality Technology and Systems, Beihang University (grant number VRLAB2019C03), and the Fundamental Research Funds for the Central Universities (grant number 328201907).

References

1. Lu, H., Zhang, Y., Li, Y., Jiang, C., Abbas, H.: User-oriented virtual mobile network resource management for vehicle communications. IEEE Trans. Intell. Transp. Syst. https://doi.org/10.1109/TITS.2020.2991766
2. Lu, H., Li, Y., Mu, S., Wang, D., Kim, H., Serikawa, S.: Motor anomaly detection for unmanned aerial vehicles using reinforcement learning. IEEE Internet Things J. **5**(4), 2315–2322 (2018). https://doi.org/10.1109/JIOT.2017.2737479
3. Lu, H., Zhang, M., Xu, X., Li, Y., Shen, H.T.: Deep fuzzy hashing network for efficient image retrieval. IEEE Trans. Fuzzy Syst. https://doi.org/10.1109/TFUZZ.2020.2984991
4. Radford, A., Metz, L., Chintala, S.: Unsupervised representation learning with deep convolutional generative adversarial networks[J]. arXiv preprint arXiv:1511.06434, (2015)
5. Karras, T., Aila, T., Laine, S., et al.: Progressive growing of gans for improved quality, stability, and variation[J]. arXiv preprint arXiv:1710.10196, (2017)
6. Karras, T., et al.: A style-based generator architecture for generative adversarial networks[C]. Proceedings of the IEEE/CVF Conference on Computer Vision and Pattern Recognition. 4401–4410 (2019)
7. Isola, P., Zhu, J.Y., Zhou, T., et al.: Image-to-image translation with conditional adversarial networks[C]. Proceedings of the IEEE conference on computer vision and pattern recognition. 1125–1134 (2017)
8. Chen, Q., et al.: Semantic human matting[C]. Proceedings of the 26th ACM international conference on Multimedia. 618–626 (2018)
9. Mao, X., Li, Q., Xie, H., Lau, R.Y., Wang, Z., Smolley, S.P.: Least squares generative adversarial networks. In: IEEE International Conference on Computer Vision (ICCV), 6 (2017)
10. Kingma, D.P., Ba, J.: Adam: a method for stochastic optimization[J]. arXiv preprint arXiv:1412.6980, (2014)
11. Simonyan, K., Zisserman, A.: Very deep convolutional networks for large-scale image recognition (2015). abs/1409.1556

Eye-Interface System Using Convolutional Neural Networks for People with Physical Disabilities

Keiichiro Kubo, Satoru Shibata, Tomonori Karita, Tomonori Yamamoto, and Shenglin Mu

Abstract In recent years, lack of effective communication method is one of the important issues that the people with physical disabilities, such as ALS (amyotrophic lateral sclerosis) face in social life. In this research we constructed an eye-gaze input system based on an image processing method using a PC and a web camera that are inexpensive and easy to install under natural light environment without the risk of illness due to the use of infrared radiation. In our proposed system, convolutional neural networks (CNN) is applied to improve the accuracy for practical use. The CNN in this study estimates the gaze position on the monitor screen from the image acquired from the web camera, and aims to obtain higher accuracy than the conventional system by learning for specific individuals.

Keywords Eye-interface · Convolutional neural networks · Communication assistance

1 Introduction

In recent years, one of the major problems for the people with disabilities in social life is the lack of effective communication measures. In many cases, although the information from caregivers can be understood by the carer-receiver, the ability to convey intentions due to physical disabilities is not efficient. In order to solve the problem, many interfaces have been developed to capture changes in human gaze direction and to input characters to a computers for communication. EOG

K. Kubo · S. Shibata · T. Yamamoto · S. Mu (✉)
Graduate School of Science and Engineering, Ehime University, Matsuyama, Ehime, Japan
e-mail: g840022x@mails.cc.ehime-u.ac.jp; shibata.satoru.mg@ehime-u.ac.jp; yamamoto.tomonori.mh@ehime-u.ac.jp; mu.shenglin.du@ehime-u.ac.jp

T. Karita
Faculty of Education, Ehime University, Matsuyama, Ehime, Japan
e-mail: karita.tomonori.mh@ehime-u.ac.jp

S. Mu et al. (eds.), *4th EAI International Conference on Robotic Sensor Networks*, EAI/Springer Innovations in Communication and Computing,
https://doi.org/10.1007/978-3-030-70451-3_7

method, scleral reflection method, corneal reflection method, etc. have been reported as useful gaze direction detection methods [1, 2]. It has been pointed out that there is a risk of illness caused by irradiating the eyeball, and that a commercially available gaze input device using infrared rays is expensive. On the other hand, some approaches have been reported to process the image of the eyeball under natural light and measure the gaze directions. Ochiai et al. employed blinking motion to perform difference image processing with opening and closing motion of eyes. The method is effective in detecting large changes in both eyes, and estimating the gaze direction [3]. The method is designed to be used under natural light sources, with no special device required, but there is a problem that the input accuracy is low due to the complicated image processing to extract the eyeball. Meanwhile, several studies have been reported on estimating gaze direction using a convolutional neural network (CNN), which is highly evaluated in the field of image recognition. Kojima et al. have proposed a method for estimating the head direction and gaze direction of a group of pedestrians [4]. The system estimates the gaze direction in an image and estimates the gaze direction based on the information. However, the gaze direction to be estimated is limited to the horizontal direction.

In light of the abovementioned problems of the conventional eye-gaze input system, we constructed a system that is inexpensive and easy to install with a PC and a web camera. In the proposed system, we designed the mechanism for the subject to use the eye-gaze motion to drive the cursor movement employing blinking motion as items selection methods. Then, to improve the accuracy of gaze direction estimation, we apply CNN to gaze input. The purpose of the CNN in this study is to determine which point on the monitor screen is being watched from the input image, and to improve the accuracy by learning.

2 Eye-Interface

Figure 1 shows the configuration of the gaze input system in this study. The user is assumed to be a patient with heavy physical disability, and a monitor screen is installed in the front direction where his/her face turns. A web camera is mounted on the monitor screen to photograph the upper body of the user. The user's face is detected from the color image obtained by the web camera. A color image of the user's upper body is acquired in real time from a web camera, a face is detected from the color image, and a range including both eyes is cut out. After a series of process the image is applied as the input of CNN.

2.1 Face Detection

As experimental conditions, the user placed the monitor and web camera in front of the subject with the distance of 600 mm as shown in Fig. 2. The subject's face can

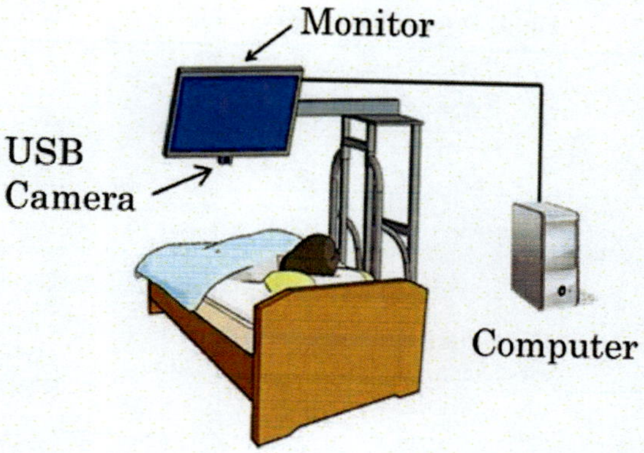

Fig. 1 The system configuration

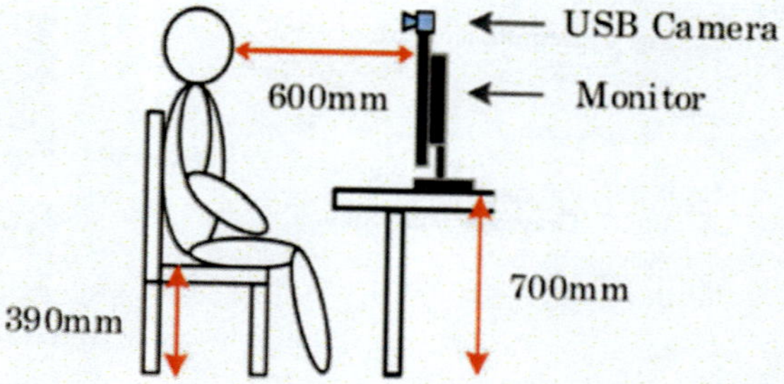

Fig. 2 Experimental condition

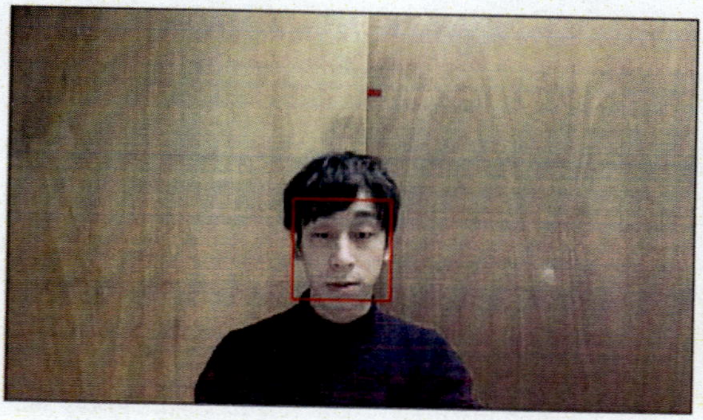

Fig. 3 Detected area in the color image

be detected from the color image (1920 pixels wide × 1080 pixels high) obtained by the web camera using the face detection function of Haar-like in OpenCV. When a face is detected in the range, a rectangular drawing (red rectangle) is performed on the face of subject, as shown in Fig. 3.

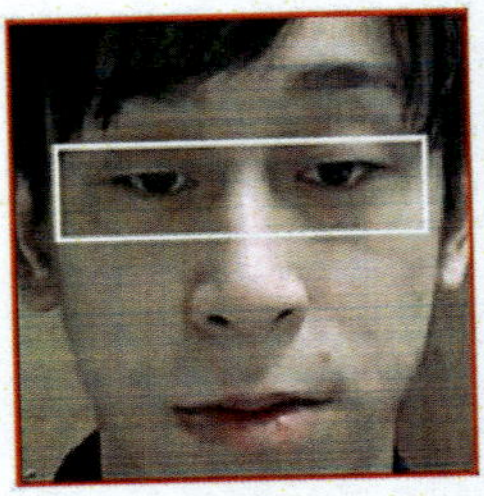

Fig. 4 Detected area of eyes

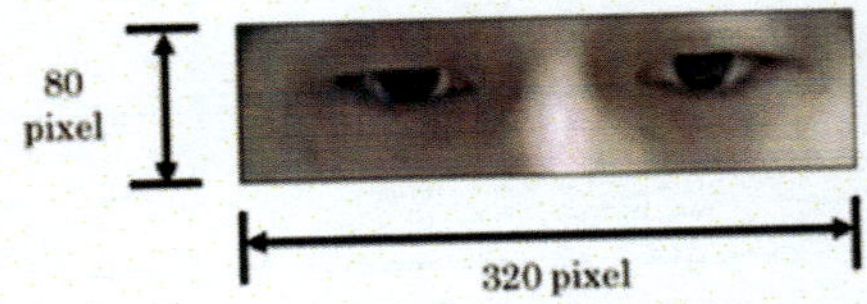

Fig. 5 Rectangular area of eyes after trimming

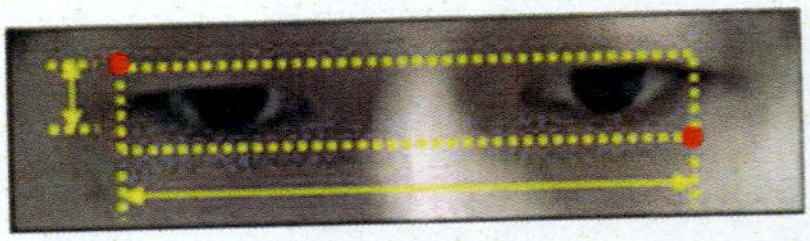

Fig. 6 Definition of area of eyes

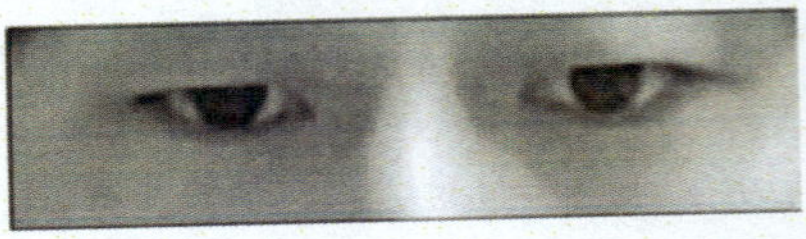

Fig. 7 Grayscale image of eyes

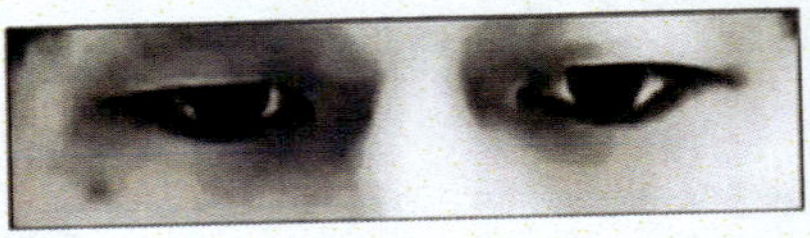

Fig. 8 Histogram equalized image of eyes

2.2 Eye Detection

With normalization on the obtained face image, both eyes include a rectangular area of 320×80 pixels drawn between two points of (40, 110) and (360, 190) which can be obtained as shown in Fig. 4. The area including both eyes is the position of the eyes shown in Fig. 5. Then, the color image of both eyes obtained is converted to a grayscale image. The results of the grayscale processing are shown in Fig. 6. The histogram equalization process is performed on the grayscale images of both eyes in Fig. 7. Detailed processing contents are shown in Eq. 1, and the results of histogram equalization processing are shown in Fig. 8.

$$g(a,b) = INT\left(\frac{\frac{S(f(a,b))}{A \times B} - min\left(\frac{S(f(a,b))}{A \times B}\right)}{1 - min\left(\frac{S(f(a,b))}{A \times B}\right)}(C-1)\right) \tag{1}$$

$f(a, b)$ and $g(a, b)$ are the value of brightness before and after the processing, respectively. $S(f(a, b))$ is the accumulative value until $f(a, b)$. $A \times B$ is the total number of pixels. C is the number of gradations of brightness (256).

3 Gaze Direction Estimation Using Convolutional Neural Network

3.1 Convolutional Neural Network

Neural network (NN) is an information processing system modeling on the mechanism of the human brain neural circuit. It can be considered as a mathematical model that has the learning ability and can automatically form the required functions based on the samples presented. NNs can be used to express nonlinear input/output characteristics by performing self-learning. Convolutional neural networks which use the conventional filtering techniques are widely used in image processing. A CNN usually consists of an input layer, a convolutional layer, a pooling layer, a fully connected layer, and an output layer. In deep learning, the convolutional layer and the pooling layer are repeated multiple times. The basic form is to form a deep layer.

3.2 Proposed Convolutional Neural Network

The structure of the CNN used in this study is shown in Fig. 9. The details of it are listed in Table 1. The input signal to the CNN is the image data obtained by the processing so far, and the output layer is assumed to be a total of two coordinate values (*x*, *y*) of the viewpoint on the monitor screen, and it is estimated where the user is looking on the monitor screen.

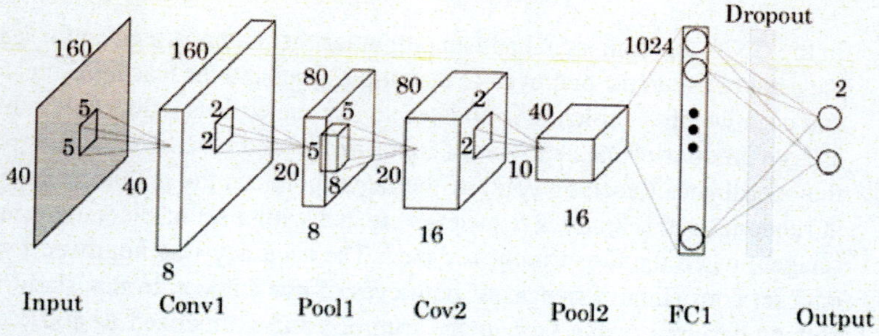

Fig. 9 Configuration of CNN

Table 1 Details of the proposed CNN

Layers	Specifications
Input layer	Image size: 160×40; Channel: 1
Convolutional layer first	Filter: 5×5×8; Activation function: ReLU; Stride: 1
Pooling layer first	Size: 2×2; Type: Maximum pooling; Stride: 2
Convolutional layer second	Filter: 5×5×16; Activation function: ReLU; Stride: 1
Pooling layer second	Size: 2×2; Type: Maximum pooling; Stride: 2
Fully-connected layer	Unit: 1024; Activation function: ReLU; Dropout: 50%
Output layer	Unit: 2; Activation function: Identity function

The configuration of the CNN used in this study is an orthodox one, and the size of the input image was 160×40 pixels. The activation function is the ReLU function, and the maximum pooling of 2 × 2 is applied. The convolution filter group in the first stage is "filter size (vertical × horizontal) × number of input layer × number of output layers" = "5 × 5 × 1 × 8." It can be expressed as a multidimensional list. Next, 5 × 5 convolution filters are added to the output of the first-stage pooling layer. In this case, the activation function is the ReLU function and a 2 × 2 maximum pooling is applied. When the second stage convolution filter is applied, eight sets of image data from eight layers can be considered as a single image data set. In the TensorFlow code, the convolution filter group in the second stage is represented by a multidimensional list of "filter size (vertical × horizontal) × number of input layers × number of output layers" = "5 × 5 × 8 × 16." Then, for the output of the second-stage pooling layer, a fully connected neural network consisting of 1024 elements is constructed. The gazing points x and y coordinates on the monitor screen are output. An output consisting of elements is obtained, and a dropout layer is provided between the nodes of the fully connected layer and the output layer, so that the connection between the nodes of the fully connected layer and the output layer is performed during learning. In this study, the dropout rate was set to 50%.

3.3 Creation of Training Data Set

In the CNN, an input signal and an output teacher signal are given for learning, and the error between the output from the output layer and the teacher signal is fed back, and each weight coefficient is updated so that the error is reduced. For this purpose, a large amount of data set (set of input image and teacher signal) is required, so the coordinates (teaching data) of the target point on the monitor screen displayed at random. In this study, a data set is created with a set of binocular images (input images) when the target point is gazed. The accuracy was improved by creating a data set consisting of images of both eyes of one subject. In this study, we use the rectangular area of the face image trimmed and normalized to 400 × 400 pixels as shown in Fig. 10. Then, as shown in Fig. 11, a rectangular area of 400 × 100

Fig. 10 Area of eyes

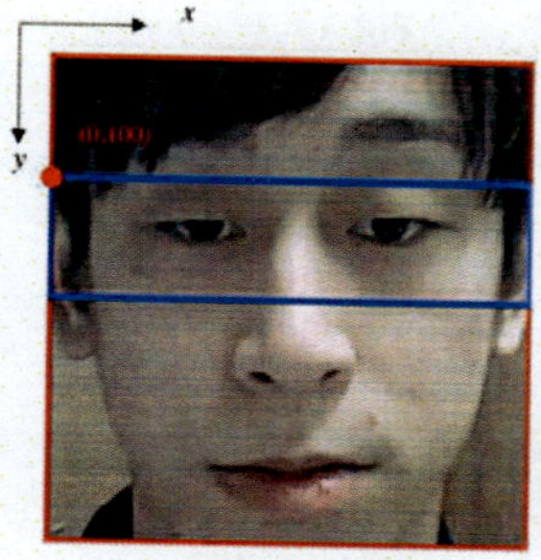

Fig. 11 Rectangular area of eyes after trimming

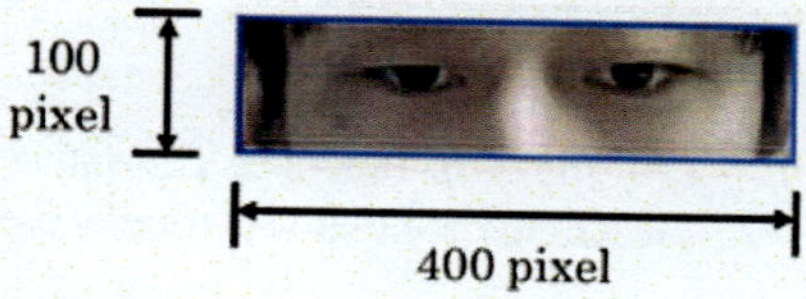

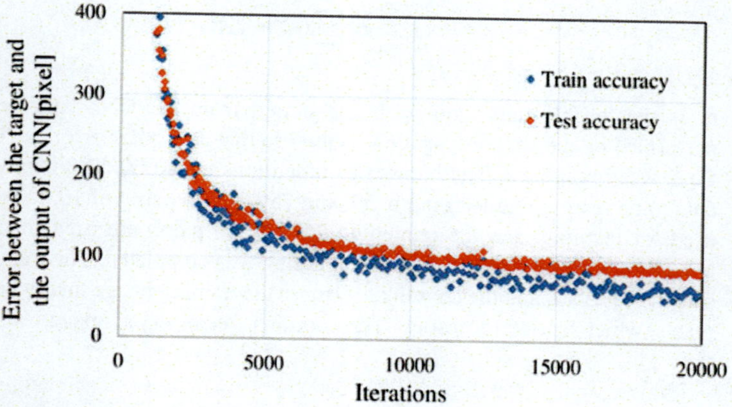

Fig. 12 Training result

pixels drawn from the two points of coordinates (0, 100) and (400, 200) is cut out. In this study, a random cropping method is applied to generate a total of 10,000 data sets, which consisted of pairs of images containing information on both eyes and coordinates on the monitor screen watched at a time.

3.4 Training of Convolutional Neural Network

Using the data set created, the CNN proposed in this study is trained, and the monitor function and the input image output from the CNN are actually observed

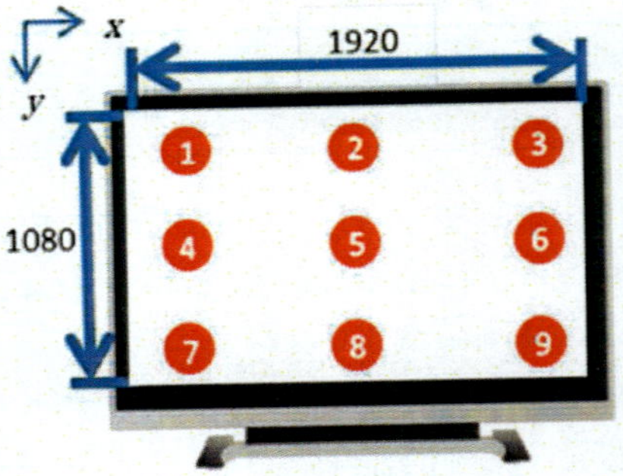

Fig. 13 Position of target points

for the error function. The square error with the monitor coordinates is used as the teaching data that was used is used. The square error is expressed as in Eq. 2.

$$E = \sum_{n=1}^{2} (r_n - y_n)^2 \quad (2)$$

r_n is the teacher data, and y_n is the output from the CNN, and learning was performed by the backpropagation method so that the square error calculated using Eq. 2 was minimized. The number of repetitions is 20,000. The number of image data to be read in one learning is 50, and 75% (7500 pairs) of 10,000 data sets are used for learning, and the remaining 25% (2500 pairs) are used for testing. The test procedure is the evaluation after learning was completed. The training result is shown in Fig. 12, and the coordinate error in the training image finally becomes 60.7 pixels, which is used for testing. The coordinate error in the image was 86.0 pixels.

4 Evaluation Experiment of Gaze Direction Estimation

With image processing procedure proposed by our group, the conventional gaze direction estimation using the feature points of the eyes (eye corners, the inner corner of the eye, and the center of the iris) is used for the evaluation comparing with the proposed CNN method. In conventional gaze direction estimation, face detection and eye detection are performed from the color image obtained by the web camera, and the outer and inner corners of the eye are detected from the binarized image obtained by automatically adjusting the threshold. Using a particle filter, a sequential tracking algorithm that simultaneously detects and tracks objects based on a set evaluation function, the center of the iris is estimated, taking into account the iris hidden by the eyelids. Where n is the number of iris contour points, m is the number of particles, and the distance between the particles and the iris contour points is d_i $(i = 1, 2, 3 \cdots n)$, the difference E_i between d_i $(i = 1, 2, 3 \cdots n)$ and the radius r of the iris is calculated using Eq. 4, and the likelihood of the particles is calculated. By using this evaluation function, particles with high likelihood, that is, points where the circle drawn with the iris radius r is closest to the iris contour point, are detected. The center point of the left and right iris, the right and left outer corners of the eye, and the x, y coordinate meters of the inner corner of the eye obtained as a result of the above processing are used as input signals. The gaze direction is estimated using a neural network with two coordinate values.

$$L_j = 1 - \frac{\sum_{i=1}^{n} E_i}{r \cdot n} \tag{3}$$

$$E_j = |d_i - r| \tag{4}$$

4.1 Experimental Setting

In the experimental design, we focused on nine target points on the monitor screen as shown in Fig. 13. The center coordinates (x, y) of the nine points are ($x_1 = 50$, $y_1 = 50$), ($x_2 = 960$, $y_2 = 50$), ($x_3 = 1870$, $y_3 = 50$), ($x_4 = 50$, $y_4 = 540$), ($x_5 = 960$, $y_5 = 540$), ($x_6 = 1870$, $y_6 = 540$), ($x_7 = 50$, $y_7 = 1030$), ($x_8 = 960$, $y_8 = 1030$), ($x_9 = 1870$, $y_9 = 1030$). The coordinate error between the target point and the gaze direction estimation result of the system are investigated.

4.2 Experimental Results

Table 2 shows the measurement results of gaze direction estimation using a conventional NN. Table 3 shows the measurement results of gaze direction estimation using the proposed CNN. Tables 2 shows that the average of the coordinate error is 155.9 pixels in the system using a conventional NN. The average of the proposed CNN's coordinate error shown in Table 3 is 83.9 pixels indicating that the system using the proposed CNN can estimate the gaze direction with high accuracy. The average difference of the coordinate errors is about twice, indicating that the system using the CNN can estimate the target point more accurately.

Table 2 Results by the conventional NN method

Points	Gazed coordinate		Output of NN		Error
	x	y	*x*	y	
1	50	50	247	100	202.8
2	960	50	1041	67	83.0
3	1870	50	1747	135	149.5
4	50	540	67	602	64.0
5	960	540	1096	699	209.7
6	1870	540	1085	551	66.2
7	50	1030	314	912	289.3
8	960	1030	1088	942	155.2
9	1870	1030	1753	888	183.5
Average					155.9

Table 3 Results by the proposed CNN method

Points	Gazed coordinate		Output of NN		Error
	x	y	*x*	y	
1	50	50	58	54	8.9
2	960	50	971	67	20.2
3	1870	50	1771	66	100.3
4	50	540	66	502	41.2
5	960	540	878	452	120.3
6	1870	540	1827	507	54.2
7	50	1030	107	1009	60.7
8	960	1030	933	1064	43.4
9	1870	1030	1565	1008	305.8
Average					83.9

5 Conclusion

In this study, we aimed to construct an inexpensive and easy-to-use gaze input interface for people with severe physical disabilities. An eye-gaze input system was constructed based on an image processing method using a commercially available PC and a web camera, which are easy to install. In order to prevent erroneous operation of the gaze selection method, we introduced a convolutional neural network that has excellent learning characteristics and can set arbitrary input/output relationships. By adopting this item selection method, the problem of erroneous input due to gaze was solved, and it was possible to operate according to the user's intention. An eye-gaze input system that is easy to operate was constructed, and feature points (eye corners, inner corners, iris center) of the eyes were input to the neural network using the image processing constructed by our conventional research group. We performed a comparative experiment on gaze direction estimation using image information including both eyes as input to a conventional NN. In the system using the network, the average coordinate error was 155.9 pixels, and in the system using the proposed CNN, the average coordinate error was 83.9 pixels. It was confirmed that the system using the CNN can track the target point more accurately.

Acknowledgments This work was supported by JSPS KAKENHI Grant-in-Aid for Scientific Research (B): Grant Number JP19H04228

References

1. Kuno, Y., Yagi, T., Fujii, I., Koga, K., Uchikawa, Y.: Development of eye-gaze input interface using EOG. J. Inf. Process. **39**(5), 1455–1462 (1998)
2. Itoh, K., Sudoh, Y., Ifukube, T.: Eye gaze communication system for people with severe physical disabilities. IEICE Trans. Inf. Syst. Pt.1 **J83-D-1**(5), 495–503 (2000)
3. Ochiai, T., Ishimatsu, T., Takami, O., Matsui, R.: Data input device for physically handicapped people operated by eye movements. Trans. Jpn. Soc. Mech. Eng. C **63**(609), 1546–1550 (1997)
4. Kojima, S., Ukita, N., Hagita, N., Yang, M.: Simultaneous temporal optimization of pedestrians' gaze orientation and gaze point in a scene. IPSJ SIG Tech. Rep. **2015-CG-161**(21), 1–8 (2015)

Proposal of Omnidirectional Movable Positioning Plate Using T-Shaped Omni Wheels

Hiroki Watanabe, Takanori Higuma, Hiroshi Narazaki, Shenglin Mu, Kanya Tanaka, and Shota Nakashima

Abstract In recent years, many positioning devices using electronic motors as a drive source have been developed. A typical example of a positioning device is the "XY stage" system, in which a nut attached to the stage moves with the rotation of a male screw. The problem with this method is that two motors, one on the X-axis and the other on the Y-axis, must be stacked to move in all directions, and the height of the device becomes high. To solve this problem, we propose a new method, the T-shaped configuration method. The omni wheels, which are arranged in a T shape, enable the plate to move in all directions without stacking the motors for the movement in *X*- and *Y*-axes. The proposed method is half the height of the conventional *XY* stage method. The principle of this mechanism is that it can move in all directions by adjusting the speed of each motor and synthesizing the speeds. In the experiments, it was verified whether the proposed T-positioning system can be used for positioning control with high accuracy under various conditions. The experimental results show that the object trajectory changed when the environment was changed, but converged to the target position, indicating that the method is suitable for various environments.

Keywords P control · Positioning control · Tracking · *XY* stage · Omni wheel

H. Watanabe (✉) · T. Higuma · H. Narazaki · S. Nakashima
Yamaguchi University, Yamaguchi, Japan
e-mail: a054vgu@yamaguchi-u.ac.jp; b058vg@yamaguchi-u.ac.jp; i059vg@yamaguchi-u.ac.jp; s-naka@yamaguchi-u.ac.jp

S. Mu
Ehime University, Matsuyama, Japan
e-mail: mu.shenglin.du@ehime-u.ac.jp

K. Tanaka
Meiji University, Kanagawa, Japan
e-mail: kanya728@meiji.ac.jp

S. Mu et al. (eds.), *4th EAI International Conference on Robotic Sensor Networks*, EAI/Springer Innovations in Communication and Computing,
https://doi.org/10.1007/978-3-030-70451-3_8

1 Introduction

In recent years, motor-driven positioning devices have been widely used in industrial equipment, such as semiconductor manufacturing equipment and machine tools. The demand for high-speed and accurate positioning is increasing year by year to improve their production efficiency. Examples of positioning control include the use of a piezoelectric device, linear motor, and *XY* stage [1–6]. In this study, we propose a solution to the *XY* stage problem. The operating principle of the *XY* stage is that the nut moves in *X*-axis direction and *Y*-axis direction when the male screw rotates. One of the problems with this device is that two motors, one on the *X*-axis and the other on the *Y*-axis, must be piled up to move in both *X* and *Y* directions, which make the equipment with a larger height. The solution to the problem of this previous research is positioning control using two omni wheels [7]. When the same motors are used, the height of the proposed method is reduced to half of that in the conventional *XY* stage because the motor is placed without stacking it. The appearance of the two omni wheels arrangement is shown in Fig. 1. The placement of the omni wheels as shown in Fig. 1 is called an opposing configuration. A positioning method using omni wheels is described below. The plate is rotated by reversing the direction's rotation of the motors on both sides. In addition, the table moves in *X* and *Y* directions by unifying the direction of rotation of the left and right motors. The operating principle of two omni wheels is shown in Fig. 2. This method has a problem that it takes a long time to move to the target position because it is moving to the target position by detouring the rotational movement. In this study, we propose an omnidirectional movable positioning plate using two omni wheels with T shape that solves the problems encountered in previous studies. This method is expected to speed up positioning control by minimizing the moving distance. Section 2 describes the T-shaped arrangement mechanism using omni wheels. Section 3 describes the positioning control using P control. Section 4 describes positioning control experiments in various environments. Finally, we conclude with a summary and prospects.

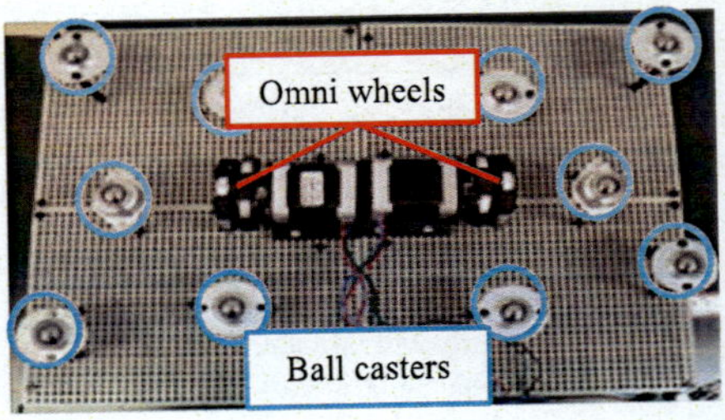

Fig. 1 Conventional arrangement

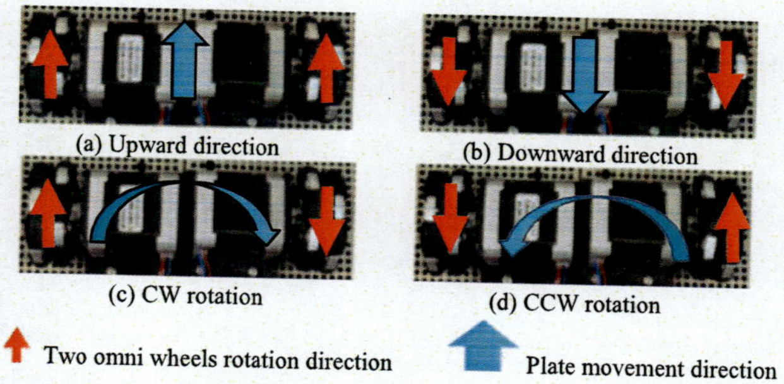

Fig. 2 Conventional operating principle. (**a**) Upward direction, (**b**) downward direction, (**c**) CW rotation, and (**d**) CCW rotation

Fig. 3 Structure of the proposed moving part

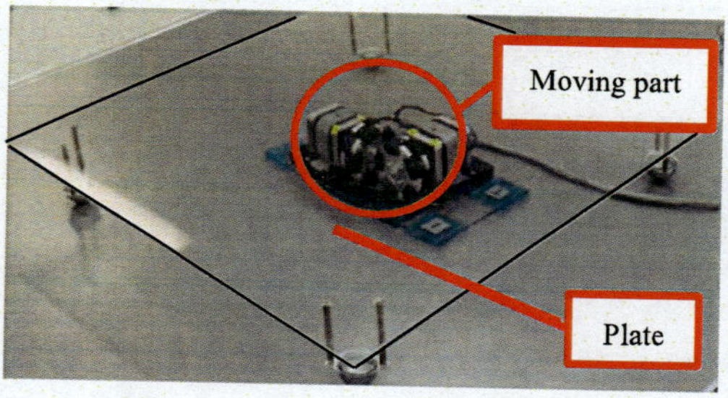

2 Plate Moving Part

2.1 Constitution

The appearance of the proposed plate moving part is shown in Fig. 3. The moving part consists of 48-mm double-aluminum omni wheels (14148, made by the NEXUS Robot) and stepping motors (ST-42BYH1004, made by MERCURY MOTOR), which is set in the center of the transparent plate. An omni wheel is a wheel that can move in various directions by the combination of the forward and backward movements of the tire with the movement of the barrel on the circumference. By using this characteristic, it is possible to move quickly even in a small space [8]. A stepping motor was used to drive the omni wheel. The stepping motor is suitable for positioning control in this study because of its high accuracy in controlling the rotation angle compared to other motors [9, 10]. The basic step of the stepping motor is 0.9°, and the distance of plate traveled by motor rotation at that angle is 0.471 mm. The speed of the motor is controlled by the motor driver (L6470H, made by STMicroelectronics). The motor driver can controls the speed at 1/128 of the angle in a single step. The minimum travel distance of the corresponding plate is 3.68 μm, which enables more precise positioning control of an object compared to the case without the motor driver.

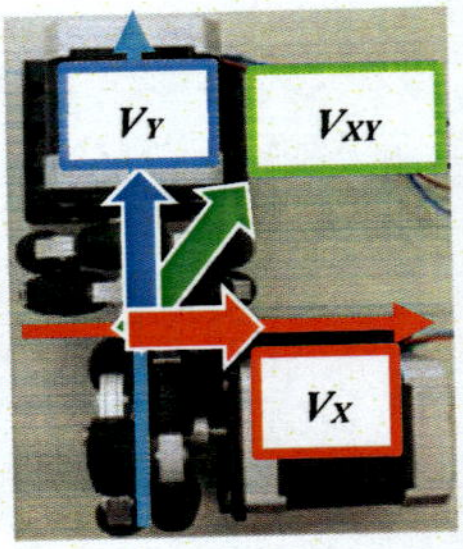

Fig. 4 Proposed omni wheels arrangement

2.2 Principle of Movement

We propose a method of moving the omni wheels in all directions over the shortest possible distance by arranging them in a T shape [11]. The principle of movement is shown in Fig. 4. The principle of movement is to synthesize the rotation speed V_X of the motor on the X-axis and the speed V_Y of the motor on the Y-axis to obtain the speed V_{XY}. Optimal control of the rotation speed and direction allows the plate to move in all directions and with the shortest distance.

3 Automatic Positioning Method Using P Control

3.1 Coordinate of an Object Acquisition Method

In this study, the color-tracking method is used to get the coordinates of the object because it is necessary to obtain the coordinates of it frame by frame. The color tracking is a method of detecting and tracking only specific colors. The color tracking method acquires the color information of the object and masks the background using the color difference between the object and the background. Recognize the unmasked location as the target and obtain the coordinates of the center of the target [12].

3.2 Positioning Method Using P Control

A camera captures the position of the object on the plate and moves the object to the target position using omni wheels. When the plate is moved, the frictional force between the plate and the omni wheels is reduced, which causes slippage. P control is one of the methods to make the measured value close to the target value and can be used to move the object at an arbitrary position to the target position in a stable manner while reducing the error due to external disturbances [13]. A block

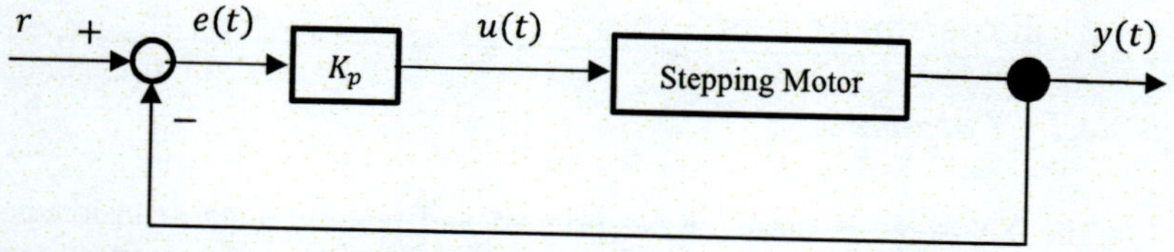

Fig. 5 Block diagram of the motor control

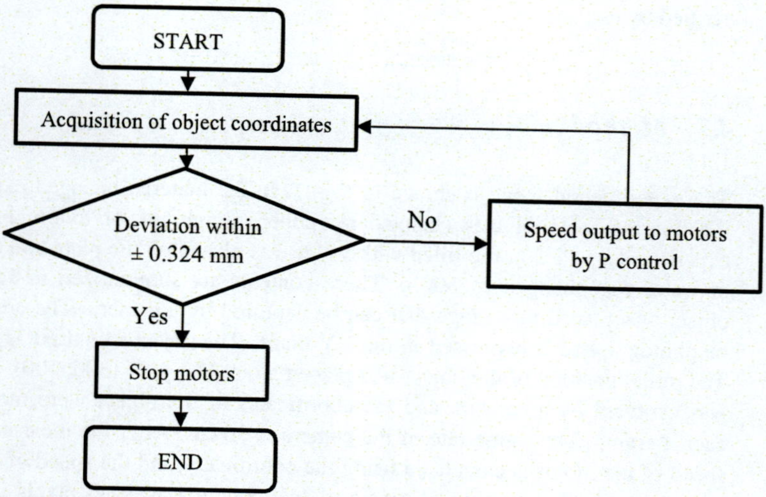

Fig. 6 Flowchart of object positioning control

diagram of the P control for the plate moving part is shown in Fig. 5. In this study, the deviation between the target position and the measured position is calculated by using the position measured by the camera as the measured value. This kind of control is called P control. By adjusting the speed of the stepping motor in response to this deviation, the position is brought closer to the target position. The block diagram of the P control for the plate moving part is shown in Fig. 6, where r is the target position (the coordinate directly below the camera), $y(t)$ is the coordinate of the object, $u(t)$ is the input (the rotation speed of the motor), $e(t)$ is the deviation, and Stepping Motor is the control target. The position of the object was represented by XY components, and the target position was approached using P control for each component. The gain factor was set at the limit of the speed at which the stepping motor did not step out.

4 Experiment

4.1 Purpose

The XY stage is used for 3D printers and requires high-precision positioning control. In addition, it must be robust to load, since it operates with various objects on the XY stage. Therefore, convergence time and trajectory were tested by varying the flooring material that affects P control. The controllability under load was also verified by us.

4.2 Method

The experimental scene is shown in Fig. 7. In the material-changed experiments, the flooring material was changed to rubber, aluminum, or nylon. In the load experiments, a PET bottle filled with water was placed on the plate and the weight of water was changed by 500 g. These contents are summarized in Table 1. An object was placed in a range that can be captured by a camera. The area that can be photographed is expressed in the XY plane. The object was used as a red ball. The initial position of the object was placed around $(x, y) = (300, 100)$. The object was captured by a camera, and the coordinates of the object were measured for each 1 frame. The frame rate of the camera is 30 fps. After the measurement, the speed of the motor is calculated using the coordinates and the speed of the motors was changed. The number of pixels of the camera was 1080 pixels in both the vertical and horizontal widths. The photographing range is 350 × 350 mm. The target position was set to $(x, y) = (540, 540)$ and the size was set to 1 × 1 pixel. The camera is placed at a height of 440 mm from the plate, and the corresponding actual distance is 1 pixel = 0.324 mm. The resolution of the camera depends on the height of the camera. The size of the plate was 500 mm in height and width.

4.3 Results

The change in trajectory with the material-changed experiments is shown in Fig. 8 and the change in trajectory with the load experiments is shown in Fig. 9. The experimental results of the change in convergence time by material are shown in Table 2 and the experimental results of the change in convergence time by the load are shown in Table 3.

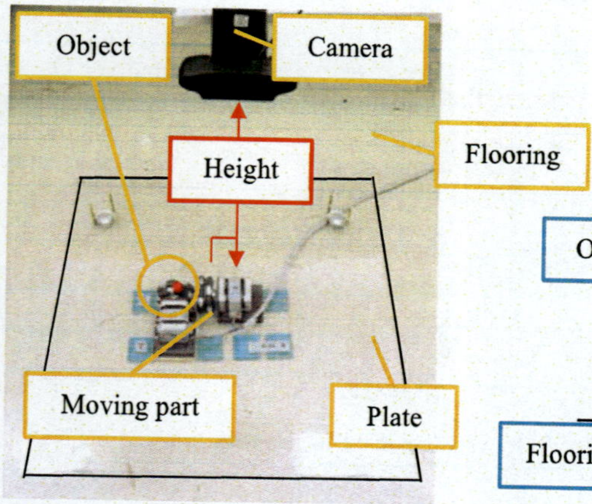

(a) Top view

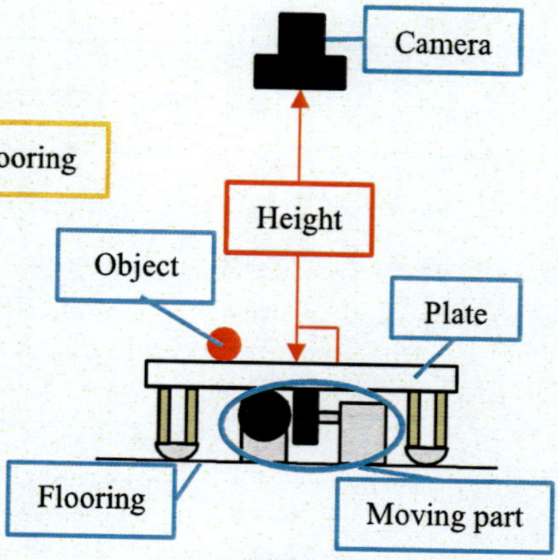

(b) Side view

Fig. 7 Experimental setting of the proposed method: (**a**) top view and (**b**) side view

Table 1 Experimental conditions

	Material-changed experiments			Load experiments			
Material	Rubber	Aluminum	Nylon	Rubber			
weight (g)	0			0	500	1000	1500

4.4 Discussion

In the material-changed experiments, the trajectory of the plate changed as the floor material was changed. The reason for the change in trajectory and convergence time is thought to be due to the change in friction between the ball casters and the ground surface. Aluminum and rubber converged on the target position almost simultaneously, while nylon converged on the target position with a significant delay. The reason for the change in convergence time is thought to be that the omni wheels and plate of aluminum and rubber did not slide when the motor speed was slowed down, while the nylon did.

In the load experiments, the trajectory of the plate changed as the load was changed. If the weight on the plate was increased, the movement of the plate became slower than when there was no load, and the plate converged at the target position more slowly as the weight became heavier. It is considered that these trajectories changed because of the friction between the plate and the omni wheels with the change of the load. The reason for the change in the convergence time is considered to be that it becomes difficult to transmit the motor motion to the plate when the load increases and it is easier to transmit the motor motion to the plate when the load decreases.

From these experiments, it was confirmed that the P control converged to the target position even if the environment was changed. It was also confirmed that the

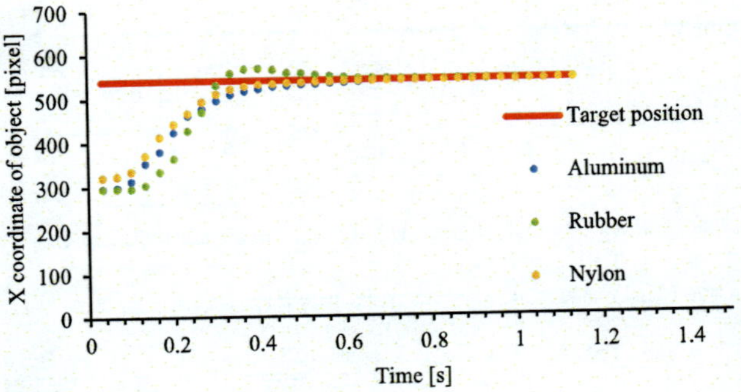

(a) Trajectory variation in the X-direction

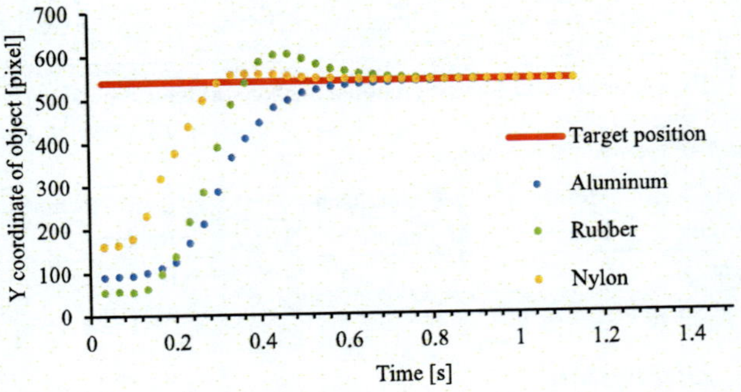

(b) Trajectory variation in the Y-direction

Fig. 8 The material-changed experiments. **(a)** Trajectory variation in the *X*-direction and **(b)** trajectory variation in the *Y*-direction

object trajectory changed and the convergence time changed when the environment was changed.

5 Conclusions

In this study, a method employing two omni wheels arranged in a T shape to move a plate in all directions is proposed, making the volume of XY stage more compact. The other is a method to automatically position an object using P control.

In the experiment, the effectiveness of the proposed moving method using P control in driving the plate to the target position is verified. We also examined the trajectory changes in various environments. The results show that the object-positioning method using P control converged to the target position. The trajectory

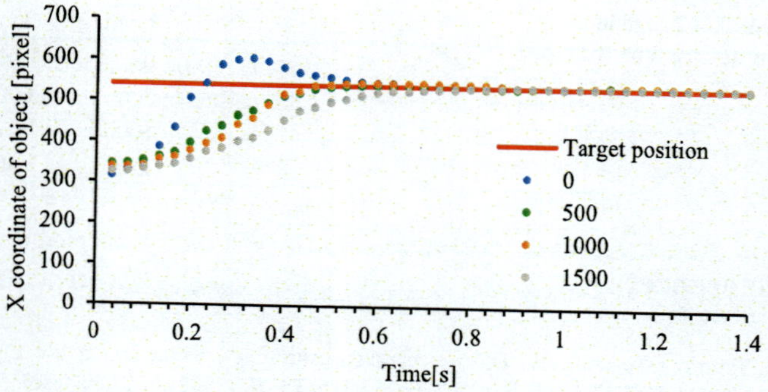

(a) Trajectory variation in the X-direction

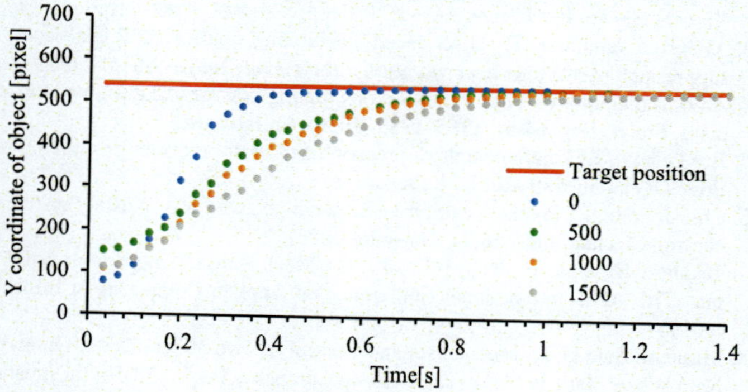

(b) Trajectory variation in the Y-direction

Fig. 9 The load experiments. (**a**) Trajectory variation in the *X*-direction and (**b**) trajectory variation in the *Y*-direction

Table 2 Variation of convergence time by material

Material	Convergence time (s)
Nylon	2.03
Aluminum	0.97
Rubber	1.00

of the plate changed due to changes in the load and the material of the flooring. The results show that the trajectory of the plate changes depending on the environment using this device, but it converges to the target position, so positioning control corresponding to various environments is possible. In future work, novel methods to optimize the variable gain are planned to be investigated [14, 15]. For safe operation of the device, a safety device is to be installed to stop the motor when the temperature of the motor exceeds a threshold value [16].

Table 3 Change in convergence time with load

Load (g)	Convergence time (s)
0	1.00
500	3.00
1000	3.27
1500	3.30

References

1. Syunro Okeguchi, F.: Ultra-precision positioning mechanism using rapid deformation of piezoelectric elements. Precis. Eng. **54**(11), 75–80 (1988). (in Japanese)
2. Masafumi Yamamoto, F., Makoto Iwasaki, S.: Analysis and improvement of positioning accuracy in ball screw-driven table system with variations of mechanical parameters. In: Kiyoshi Ohishi. IEEJ Trans, IA, val. 128, pp. 839–848. (2008). (in Japanese)
3. Akihiro Yamamoto, F.: High speed positioning control of linear motor drive table with suppression of machine base vibration. Precis. Eng. **70**(5), 645–650 (2004) (in Japanese)
4. Akio Yamamoto, F.: High precision positioning control using high-power electrostatic linear drive. Precis. Eng. **64**(9), 1385–1389 (1998). (in Japanese)
5. Jiro Otsuku, F.: Fundamentals of precision and ultra-precision positioning technology. Precis. Eng. **77**(7), 656–659 (2011). (in Japanese)
6. Chih-Jer Lin, F.: Precise positioning of piezo-actuated stages using hysteresis-observer based control. Mechatronics. **16**(7), 417–426 (2006)
7. Takanori Higuma, F., Kanya Tanaka, S.: Meal assistant robot with omnidirectional mobile plate. 7th ACIS International Conference on Applied Computing & Information Technology (2019)
8. Masanori Sakaki, F.: Mechanism and Control of Two-Wheel Drive, Omni-Wheel and Mecha-Nam Wheel, 1st edn. An Impress Group Company, Tokyo (2019). (in Japanese)
9. Makabe Kuniaki, F.: Control Circuit Design for Stepping Motors, 4th edn. CQ Publishing Co., Ltd, Tokyo (1990). (in Japanese)
10. Takashi Kenjo, F., Yoshiki Nimura, S.: Control Circuit Design for Stepping Motors, 10th edn. General Electronic Publishing Company, Tokyo (1988). (in Japanese)
11. Honda Homepage. https://global.honda/innovation/robotics/UNI-CUB.html. Accessed 16 Oct 2020
12. Masanao Koeda, F., Etsuko Ueda, S., Takayuki Nakamura, T.: Introduction to Image Processing with OpenCV, 2nd edn. Kodansha Ltd., Tokyo (2017). (in Japanese)
13. Yuki Minami, F.: PID control (basic edition, <Feature>Informative guide to control engineering for beginners). System/Control/Information. **56**(4), 192–195 (2012). (in Japanese)
14. Mehdi Nasri, F.: A PSO-based optimum design of PID controller for a linear brushless DC motor. Eng. Technol. Int. J. Electr. Inf. Eng. **1**(2), 171–175 (2007)
15. Huimin Lu, F., Ming Zhang, S.: Deep fuzzy hashing network for efficient image retrieval. IEEE Trans. Fuzzy Syst. (2020). https://doi.org/10.1109/TFUZZ.2020.2984991
16. Huimin Lu, F., Yujie Li, S.: Motor anomaly detection for unmanned aerial vehicles using reinforcement learning. IEEE Internet Things J. **5**(4), 2315–2322 (2018)

A General Pseudo-Random Number Generator Based on Chaos

Jianwen Lv, Xiaodong Li, Tao Yang, Haoyang Yu, and Beisheng Liu

Abstract In view of the shortcomings and problems of the commonly used C language pseudo-random number generation function, the existing C language pseudo-random number generation function is improved, a general pseudo-random number generator based on chaos is proposed, and multiple types of random number acquisition interface are introduced. While improving the randomness of the original C language pseudo-random number function, the procedure of improvement also enhances its versatility and can better meet the needs of different types of random numbers. The test results of the pseudo-random number generator show that the random number generated by it has good randomness, and the function call is convenient and flexible.

Keywords Chaos · Pseudo-random number generator · Random sequence retrieval · Universality

1 Introduction

Pseudo-random number generator (PRNG) is widely used in various fields such as system simulation and security [1]. Based on a reliable and efficient pseudo-random number generator, the system's operation, evolution, and development process are truly described in the system simulation. In the field of information security based on cryptography, pseudo-random number generators also play an important role. Key generation, digital signatures, authentication and identification, and various secure communication protocols are inseparable from high-quality random numbers. In a

J. Lv · X. Li (✉) · H. Yu · B. Liu
Beijing Electronic Science and Technology Institute, Beijing, China

T. Yang (✉)
The Third Research Institute of Ministry of Public Security, Shanghai, China
e-mail: yangtao@stars.org.cn

S. Mu et al. (eds.), *4th EAI International Conference on Robotic Sensor Networks*, EAI/Springer Innovations in Communication and Computing,
https://doi.org/10.1007/978-3-030-70451-3_9

sense, the security of random numbers determines the security of the entire security system [2].

There are four commonly used pseudo-random number generation functions in the C language standard library <stdlib.h> – rand function, srand function, randomize function, and random function. But these functions have obvious defects and poor ease of use. This chapter analyzes the existing pseudo-random number generation function in C language, proposes a general pseudo-random number generator scheme based on chaos, and designs a more general function interface. Compared with the original random number generation algorithm in C language, the pseudo-random number generator has enhanced anti-cracking ability and can generate pseudo-random numbers of different lengths, different ranges, and different types according to the needs of users. In addition, a key string is generated after the pseudo-random sequence is generated, which can be used to retrieve the lost pseudo-random number sequence under special circumstances. The test results show that the method proposed in this chapter has good performance in terms of randomness, key sensitivity, and compatibility.

2 Analysis of Commonly Used Pseudo-Random Number Generating Functions

2.1 *Rand Function*

The principle of the rand function in generating a pseudo-random number is to read a number as the seed parameter of the function. This parameter is determined after the computer is turned on. The function performs initialization operations based on the seed parameter and generates a pseudo-random sequence by iteration.

2.2 *Srand Function*

The principle of the srand function in generating pseudo-random numbers is the same as the rand unction, however, to make up for the shortcoming that the rand function generates the same pseudo-random sequence every time it is turned on, the srand function provides the function of customizing seed parameters.

2.3 *Randomize Function*

The randomize function is a random number generator initialization function, which is equivalent to a function that can modify the system seed parameters. Therefore, as long as the randomize function is called before the rand function to modify the system seed parameters, it can also generate a non-repetitive pseudo-random number sequence. However, the highest accuracy of the system time it obtains is still seconds, which is also vulnerable to exhaustive cracking attacks.

2.4 *Random Function*

Although the random function is similar to the rand function in usage, it can limit the pseudo-random number generation range by setting the value of formal parameters [4]. This function limits the value of the pseudo-random sequence by taking the remainder of the seed parameter iteration, but its shortcomings are also very obvious, that is, the minimum value of the range of values cannot be limited and the minimum value can only be 0.

3 One-Dimensional Logistic Map

Due to the ergodic, random, and initial value and parameter sensitivity characteristics of chaotic maps, chaotic sequences generated using this property have good cryptographic properties [3].

The one-dimensional logistic map is a very simple chaotic map from the mathematical form. As early as the 1950s, many ecologists used this simple difference equation to describe the change of population [5]. This system has extremely complex dynamic behavior and is widely used in the field of secure communications. Its mathematical expression is as follows:

$$f(x) = \mu x\,(1 - x)\,, \quad x \in [0, 1] \tag{1}$$

Among them, $\mu \in [0,4]$ is called logistic parameter. Research shows that when $x \in [0,1]$, the logistic mapping function is in a chaotic state [6].

The chaotic system is extremely sensitive to the initial value, the difference between the initial value is very small, and the difference between the values after several iterations is extremely large. For example, when $\mu = 4$ and the initial value $x(0) = 0.3$, after 20 iterations, $x(20) = 0.3794$. When the initial values differ by only 0.00001, that is, $x(0) = 0.300001$, $x(20) = 0.0084$, and after 20 iterations, $x(20) = 0.0084$. The iteration result differs by 45 times. The results of the two iterations are shown in Figs. 1 and 2. The extremely sensitive nature of the initial value determines that the time series generated by the chaotic system is highly unpredictable [7].

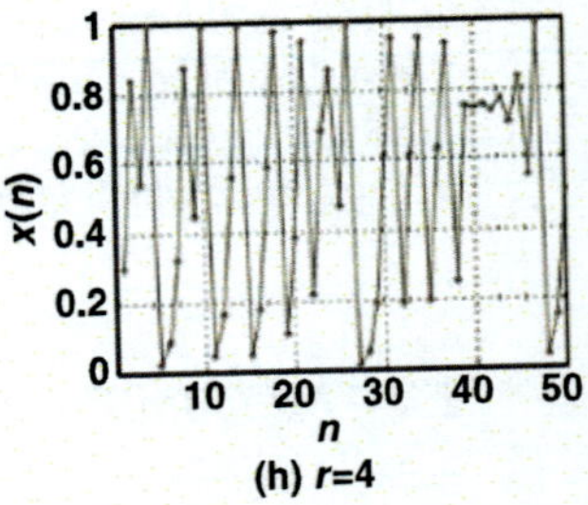

Fig. 1 *x* iteration result with initial value of 0.3

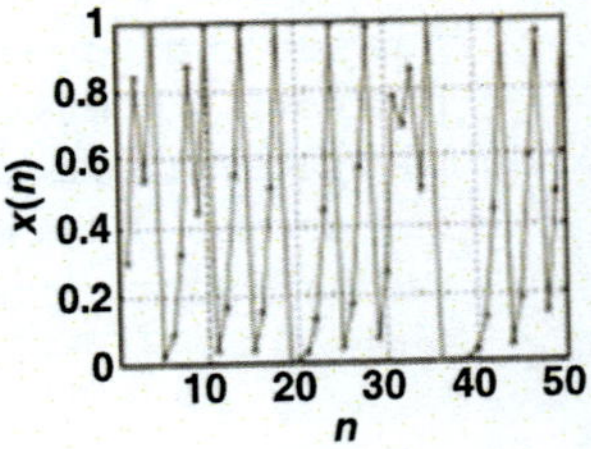

Fig. 2 *x* iteration result with initial value of 0.300001

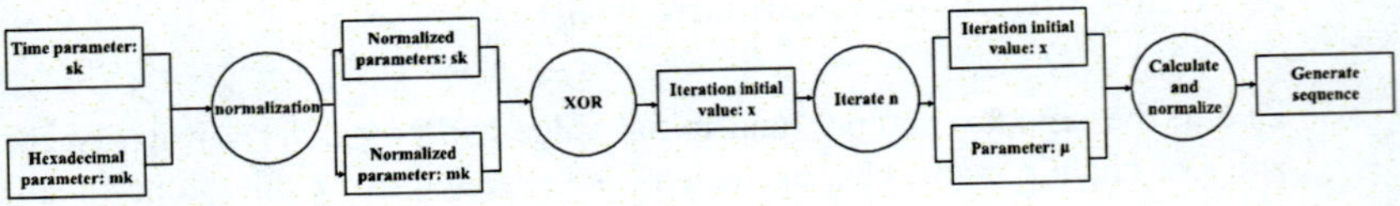

Fig. 3 Algorithm workflow

4 Algorithm Design Principles

For the two initial parameters μ and x of the traditional one-dimensional logistic mapping function, only when their values are within the specified interval, the sequence generated by the logistic mapping function is non-periodic and non-convergent. In addition, the generated sequence must converge to a certain value [5] (Fig. 3).

This chapter proposes not to directly generate the entire pseudo-random number sequence, but to generate a single pseudo-random number at a time through grouping iterations, etc., to effectively improve the randomness of the generated sequence. At the same time, the two parameters of this improved one-dimensional logistic mapping function can be determined by multiple factors. This system uses the results of four sets of parameter iterations as the coefficient μ of the system. The initial value x of the chaotic sequence is determined by three parameters mk, sk, and n, and they are defined as follows:

- μ: This parameter is four sets of system parameters used for iteration, and its range is $(0 \leq \mu \leq 4)$.
- sk: The parameter sk is obtained by multiplying the system parameters year, month, day, hour, minute, second, and microsecond in this scheme.

mk: The parameter mk is a set of four fixed hexadecimal parameters. Each time, one of the four sets of parameters is used to iterate in order to prevent the problem of randomness caused by using only one parameter.

n: The parameter n represents the number of system pre-iterations before generating chaotic sequences. By increasing the number of iterations, a more random chaotic sequence can be generated.

5 General Pseudo-Random Generation Interface Design

A good pseudo-random number generator needs to take into account both randomness and versatility. Randomness can be tested with NIST and other related testing software. At present, the randomness of most pseudo-random number generators can pass the test, but the versatility is not strong. In order to improve the versatility, this chapter proposes a set of pseudo-random number generator functions and provides an external interface to facilitate other programs to call. The required pseudo-random sequence can be obtained by entering the relevant parameters.

Different functions of the interface are implemented in different interface functions. This chapter provides three basic types of random number generation interfaces, which can generate corresponding integer pseudo-random numbers, floating-point pseudo-random numbers, and character pseudo-random sequences. The corresponding functions are: hdintrand function, hddoublerand function, hdcharrand function.

6 Performance Test

Regarding the testing of random/pseudorandom sequences, the NIST (National Institute of Standards and Technology) provides 16 detection indicators [8]. For each binary sequence, each test indicator will give a P-value as the test result. If the value is greater than a pre-set threshold α, it means that the randomness of the test sequence is 1-α, or the sequence has passed the randomness test of the detection index; otherwise, it means that the index has not passed test. In this paper, α is set to 0.01, that is, if the sequence passes the test, the credibility of its randomness is 99%.

At the same time, 100 pseudo-random sequences were generated using the algorithm and random module based on the work described in this chapter, and each sequence has a length of 100,000 bits. These sequences are applied in randomness tests, the P- value of each detection index is shown in Table 1. Based on randomness test, the results generated by the pseudo-random number generator in this paper are better than the random function test results overall. Based on the comprehensive experimental results, it can be considered that the pseudo-random number generator

Table 1 Randomness test results according to NIST indicators

Test item	P-value generated by our algorithm	Random number generated by random module
(Frequency)	0.017912	0.920402
(Block frequency) ($m = 128$)	0.107293	0.038105
(CumulativeSums) – forward	0.242986	0.855781
(CumulativeSums) – reverse	0.186566	0.929658
(Runs)	0.689019	0.362766
(LongestRunofOnes)	0.392456	0.237054
(Rank)	0.105618	0.329090
(FFT)	0.186566	0.912193
NonOverlappingTemplate	0.392456	0.531696
(OverlappingTemplate) ($m = 9$)	0.141256	0.136782
(Universal)	0.689019	0.439164
(ApproximateEntropy) ($m = 10$)	0.311542	0.826978
(RandomExcursions) ($x = +1$)	0.275709	0.152625
(RandomExcursionsVariant) ($x = -1$)	0.162606	0.496248
(Serial) ($m = 16$)	0.141256	0.161351
(LinearComplexity) ($M = 500$)	0.179129	0.088454

proposed in this chapter can generate pseudo-random number sequences with good cryptographic characteristics [8].

7 Conclusion

In this chapter, a C-language universal pseudo-random number generator based on chaotic system is designed. The algorithm of this pseudo-random number generator is described in detail, its versatility and security are introduced, and its chaotic characteristics are analyzed. At the same time, this chapter analyzes some defects of the existing pseudo-random number generation function in C language. The comparison test results show that the random number generator design scheme given in this chapter has good random performance and strong versatility. According to the random sequence and other characteristics, the pseudo-random sequence generated in this study is enough to meet the randomness requirements.

Acknowledgments This work was partially supported by the National Natural Science Foundation of China (grant number 62072014), the Beijing Natural Science Foundation (grant number L192040), the Open Project Program of State Key Laboratory of Virtual Reality Technology and Systems, Beihang University (grant number VRLAB2019C03), and the Key Lab of Information Network Security, Ministry of Public Security (grant number C18612).

References

1. Liu, P.H., Lu, H.X., Gong, G.L., et al.: High-speed arbitrary distributed pseudo-random number generator based on FPGA[J]. J. Appl. Sci. **30**(3), 306–310 (2012) (in Chinese)
2. Su G.P.: Research on random sequence and application of wavelet analysis in information security[D]. Graduate School of Chinese Academy of Sciences (Institute of Electronics), 2002 (in Chinese)
3. Zhang, X.F., Fan, J.L.: A new piecewise nonlinear chaotic map and its performance[J]. Acta Phys. Sin. **59**(04), 2298–2304 (2010) (in Chinese)
4. Li, K.J.: On the random number problem in C language[J]. J. Changchun Univ. **6**, 64–68 (2008) (in Chinese)
5. Zhu, Y.P.: Performance comparison research of various chaotic systems[J]. Microcomput. Appl. **35**(12), 4–6 (2016) +9. (in Chinese)
6. Zhao, G., Tian, Y.L., Yin, S.: Progresses and perspectives of Chaotic Cipher algorithm[J]. J. Beijing Inst. Electron. Technol. **24**(04), 9–14 (2016) (in Chinese)
7. Sun, X.H., Lin, Q.H., He, Y.W.: Pseudorandom number generator based on combined chaotic map[J]. Chin. J. Sci. Instrum. (S1), 805–807 (2006) (in Chinese)
8. Zhang, Y.Q., Li, S.B., Qu, S., Lv, K.L., Liu, C., Xu, X.R.: NIST randomness test method and application[J]. Comput. Knowl. Technol. **10**(26), 6064–6066 (2014)

A Light Chaotic Encryption Algorithm for Real-Time Video Encryption

Beisheng Liu, Xiaodong Li, Haoyang Yu, and Jianwen Lv

Abstract With the increasing application of video, security problems caused by video leakage also become more prominent. Due to the randomness, ergodicity, determinism, and sensitivity to initial conditions of chaotic systems, it is possible to generate a large number of uncorrelated pseudo-random chaotic sequences, so chaotic encryption has become an important research direction for video security. This chapter improves on the basis of chaotic encryption algorithm based on logistic mapping and proposes an optimized chaotic encryption algorithm. The algorithm introduces tent mapping, which complements the logistic mapping to achieve superiority, thereby obtaining better performance than the two algorithms. On this basis, a prototype system for real-time video transmission was designed and implemented to verify the feasibility of the encryption algorithm.

Keywords Video encryption · Chaos · Randomness

1 Introduction

With the development of multimedia and computer network technology, the scope of multimedia video applications has expanded to many fields such as economy, military, politics, and education. The confidentiality and security requirements in sensitive areas such as political, economic, and military challenge the efficient and secure encryption algorithms for video. Therefore, the research on the security of video has become a hot topic in recent years [1].

British scholar Robert A.J. Matthews first explicitly proposed the "chaotic cipher" in 1989 and designed a chaotic sequence cipher scheme based on logistic mapping and made a certain deformation. Due to the fundamental characteristics of chaotic systems such as randomness, ergodicity, determinism, and sensitivity to initial conditions. Therefore, chaotic encryption has become an important direction in the research of video encryption algorithms [2, 6].

B. Liu · X. Li (✉) · H. Yu · J. Lv
Beijing Electronic Science and Technology Institute, Beijing, China

S. Mu et al. (eds.), *4th EAI International Conference on Robotic Sensor Networks*, EAI/Springer Innovations in Communication and Computing,
https://doi.org/10.1007/978-3-030-70451-3_10

Based on the study of the chaotic algorithm, this paper builds a real-time video transmission prototype, improves the chaotic encryption algorithm based on logistic mapping, and performs randomness test and analysis on the generated chaotic sequence [3], and the algorithm can securely and quickly encrypt video in real time.

2 Design and Construction of Video Transmission Prototype System

A prototype video transmission system including video capture, video transmission, and video playback functions was designed and implemented.

Server Side:

Use OpenCV library in QT to realize operations such as capturing video and capturing video frames with notebook camera. Set the frame rate of the video frame collection by yourself and encode the collected video frames. Then use the JRTPLIB library to create an RTP session, divide a video frame into multiple data packets, and send it to the client through the RTP protocol to ensure that the client plays the video stream at a normal speed.

Client Side:

Like the server, create a thread, use the JRTPLIB library to create an RTP session to receive the data packet, decode it when a frame of data is completely received, and then display the video frame with a picture display control.

The design block diagram of the video transmission prototype is shown in Fig. 1.

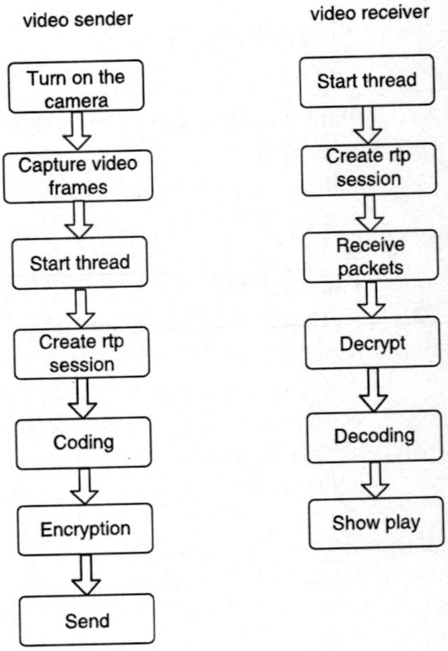

Fig. 1 Video transmission prototype design block diagram

3 Design and Implementation of Chaotic Encryption Algorithm

The chaotic encryption algorithm in this paper is based on logistic mapping and is proposed along with the concept of "chaotic encryption" [4, 5]. As a chaotic algorithm with a long history, logistic mapping is very simple and widely used. We have made some optimizations and improvements in key generation, key handling, and encryption methods to make the encryption algorithm more secure and efficient.

3.1 Design of Encryption Algorithm

The significance of our research is to design an algorithm suitable for real-time video encryption, which requires high efficiency of the algorithm.

The chaotic encryption algorithm is based on logistic and one-dimensional mapping. The algorithm is relatively simple, with high iteration speed. However, it has complex dynamic behavior, is sensitive to the initial value. The generated sequence of the algorithm is a pseudo-random sequence, which meets the requirements of encrypted real-time video.

Chaotic sequence is a mixture of disorder and order. In order to solve the problem that the sequence iterated at the beginning of the logistic map may not be random enough, the sequence generated in the previous iteration can be run empty for a while, and the multi-way sequence can be used to cover, so that the sequence generated for encryption is basically disordered.

The logistic algorithm is based on an initial value and then iteratively generates a sequence. We start with the generation of the initial value. The iterative initial value can be generated by a master key and a set of one-time keys, where the master key is fixed and the one-time key is randomly generated. The initial key is randomly generated, and the XOR is selected to process the key to ensure the balance of "0" and "1". When encrypting with the plain text, XOR encryption is also selected. The one-time key consists of four 32-bit integers, and each number has a value between 0 and 0xffffffff. In order to make it meet the requirements of logistic mapping (between 0 and 1), generally the number is continuously divided by 2, but due to the numerical specificity of the number selected in this scheme, each number can be divided twice with 0x10000 to get the initial value quickly.

The final part is the encryption part. In this paper, four-way coverage is selected, i.e., the four sequences are XORed and then encrypted with the plaintext XOR. One byte is encrypted once and every bit is encrypted, so only one byte is needed for encryption at a time. In summary, the flowchart of chaotic encryption algorithm based on logistic mapping is shown in Fig. 2:

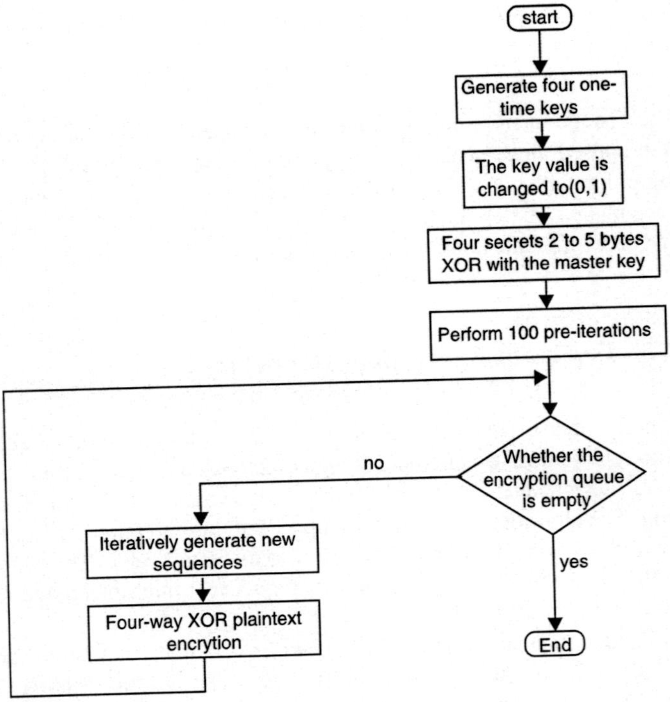

Fig. 2 Encryption algorithm program block diagram

3.2 *Implementation of Encryption Algorithm*

The encryption algorithm has the following process:

1. We first use the time function to generate four 8-bit hexadecimal numbers, i.e., the value is between 0x00000000 and 0xffffffff, so each number can be changed to (0,1) after two times of division with 0x10000, which greatly saves the time to generate the initial key, and can generate four keys at a time. In this paper, we first use the time function to generate four 8-bit hexadecimal numbers, i.e., the value is between 0x00000000 and 0xffffffff, so each number can be changed to (0,1) after two times of division with 0x10000, which greatly saves the time to generate the initial key, and can generate four keys at a time.
2. The generated one-time key is processed, instead of using all the bytes of the generated key, only the second to fifth bytes are used, and the four bytes are XORed with the master key to generate a new key, and then the new key is used

to iteratively generate chaotic sequences. In this way, the double guarantee of the master key and the one-time key makes it more difficult to crack. At the same time, let the newly generated sequence run for a while, remove the less random part in front, and increase the reliability of the sequence.

3. The part encrypted with the plaintext uses XOR operation. Combine a plaintext byte with the fifth byte of the first key, the fourth byte of the second key, the third byte of the third key, and the second of the fourth key XOR (four-way XOR coverage).

3.3 Improvement of the Encryption Algorithm

For better security and performance, the following improvements are adopted to our encryption algorithm:

1. Improvement of parameter μ

For logistic mapping, the two main factors that affect the sequence are the initial value and the other is μ. The change of μ has a great influence on the blank window of the sequence and whether there is a blank area. We found that $\mu = 4$ is the most ideal state, but this is a fixed value, which is not realistic for the encryption algorithm, and so we can only think of ways to make the value of μ as close to 4 as possible.

In the original algorithm, the value of μ is fixed, and the changed algorithm dynamically changes the value of μ. After each iteration, it is determined whether the value of μ is greater than 4, and if it is greater than 4, the value of μ minus the key value is divided by 10, i.e., a number between 0 and 0.1 is subtracted. If the value of μ is less than 4, divide μ plus the key value by 100, and in this cycle, control the value of μ around 4. The result of the assumption is that even if the blank serial port and blank interval cannot be completely avoided, the value of μ can be relatively reduced, and the value of μ changes every time a new sequence number is generated, which is more difficult to crack.

2. The mixture of tent and logistic mapping

The flaw of logistic is that the sequence values are unevenly distributed and there are blank windows, and the flaw of tent mapping is that it is easy to fall into the fixed points and small cycles. When the tent map falls into a small period or a fixed point, it is initialized using logistic mapping. When to fall into a small period and fixed points is difficult to judge in the encryption process, so in this topic, I tried to use the logistic map to initialize the tent map every 50, 100, 200, 500, 1000 times, and the sequences generated by each algorithm were tested.

4 Implementation and Test of Video Encryption System Based on Chaos

4.1 Realization of Video Encryption System Based on Chaos

According to the design of the previous video transmission system, the encrypted part should be placed after encoding to encrypt the jpg image data, and then decrypted and then decoded after receiving at the other end. The situation of the entire system is shown in Fig. 1. The combination of chaotic algorithm and video transmission system should be slightly modified on the basis of the original algorithm. Mainly in the storage and transmission of the one-time key, the video collection end will use the time function to generate a set of one-time key when encrypting the video. This key is needed for decryption. The one-time key and the video frame data are sent to the video receiving together. The key is put at the front of the first data packet, and it will be extracted first in receiving procedure.

4.2 System Test

The main purpose of test is to confirm the connections, transmission in the sending and receiving procedures between two computers. The result shows that it can be connected normally, indicating that both the network function and the encryption and decryption functions are realized, and the videos on both ends are basically of same smoothness, indicating that the encryption speed is still OK. The results are shown in Figs. 3 and 4.

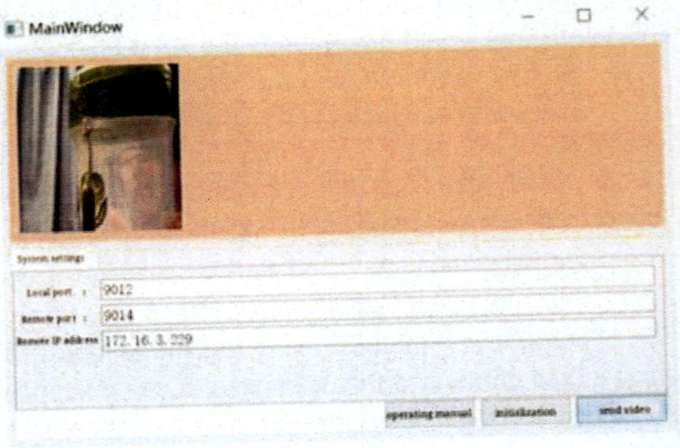

Fig. 3 Video capture terminal

Fig. 4 Video receiver

5 Conclusion

The research work discussed in this chapter designs and implements a real-time video transmission system based on chaotic encryption, and analyzes and improves the encryption algorithm. The core of this system research is in the encryption and decryption part. The result showed that the proposed algorithm based on logistic mapping is an efficient and secure encryption one for real-time video.

In view of the defects of logistic mapping, this research proposes two improvement schemes. We dynamically change the value of the parameter μ so that it keeps getting closer to 4, and when it is greater than or equal to 4, it returns to a value close to 4 and continues to cycle; then, it is mixed with another chaotic mapping function tent map. The sequence values of tent map are evenly distributed, but it is easy to produce small periods and fixed points after a certain number of iterations, so the two are combined to complement each other. After a certain number of iterations of the tent mapping, the logistic mapping is used to initialize it.

After testing, it is found that these two improved schemes have improved the randomness of the sequence and improved the security and reliability of the algorithm. Finally, the improved algorithm is also used in the real-time video transmission system and it can smoothly realize the encryption and decryption of the real-time video.

Acknowledgments This work was partially supported by the National Natural Science Foundation of China (grant number 62072014), the Beijing Natural Science Foundation (grant number L192040), and the Open Project Program of State Key Laboratory of Virtual Reality Technology and Systems, Beihang University (grant number VRLAB2019C03).

References

1. Hu Guojie, Feng Zhengjin: Security property of a class of digital chaotic encryption system[J]. J. Electron. Inf. **25**(11), 1514–1518 (2003)
2. Li En, Wu Min, Xiong Yonghua: Design and application of encryption algorithm based on double chaos map[J]. Appl. Res. Comput. **26**(4), 1512–1514 (2009)
3. Frank Dachselt, Wolfgang Schwarz: Chaos and cryptography[J]. IEEE Circuits Syst. I Basic Theory Appl. **48**(12), 1498–1509 (2001)
4. Wang, L., Wang, F., Wang, Z.: Novel Chaos-based pseudorandom number generator[J]. Acta Phys. Sin. **55**(8), 3964–3968 (2006)
5. Fu, C., Chen, J.J., Zhou, H.: A Chaos-based digital image encryption scheme with an improved diffusion strategy[J]. Opt. Express. **20**(3), 2363–2378 (2012)
6. Xiaodong Li, Haoyang Yu, Hongyu Zhang, Xin Jin, Jing Liu: Video encryption based on hyperchaotic system. Multimedia Tools Appl. **79**, 23995–24011 (2020). https://doi.org/10.1007/s11042-020-09200-1

Intelligent Analysis and Presentation of IOT Image Collection in Private Cloud

Hongyu Zhang, Chaoen Xiao, Xiaodong Li, and Beisheng Liu

Abstract More and more people find it tedious to search the photo they need in such a large number of photos. Therefore, it becomes more and more important to develop efficient and feasible methods to help users retrieve, browse, or automatically label their albums. With the development of computer intelligence, it is possible for the computer to analyze and express images in IOT. Although there are some album management systems and schemes for analyzing and representing images, they have poor security, weak computing power, and high coupling. To make up for these shortcomings, this chapter proposes a multi-terminal smart album management scheme based on edge computing and private cloud. To protect user data and personal privacy, we build private cloud storage to store images by NAS technology. At the same time, in order to improve the intelligent level and aesthetic ability of album management system, our scheme also integrates edge computing to support intelligent algorithms with high computational power consumption. Based on microservices, the scheme decouples the functions of front-end and back-end, and provides users with convenient multi-end access services for different scenarios.

Keywords Intelligent album · Image collection · Private cloud · Edge computing · Micro-services

1 Introduction

More and more people find themselves taking thousands of photos, but their photo management tools do not provide perfect functions for effective management of these photos. It is a very complicated thing to look for the photos we need in such a large number of photos. Therefore, it becomes more and more important to develop efficient and feasible methods to help users retrieve, browse, or automatically label

H. Zhang · C. Xiao (✉) · X. Li · B. Liu
Beijing Electronic Science and Technology Institute, Beijing, China
e-mail: xce@besti.edu.cn

S. Mu et al. (eds.), *4th EAI International Conference on Robotic Sensor Networks*, EAI/Springer Innovations in Communication and Computing,
https://doi.org/10.1007/978-3-030-70451-3_11

their albums. With the development of computer intelligence, it is possible for the computer to analyze and present images in IOT. The application development of image software with artificial intelligence technology has become a very important and popular field in software development.

Now many album management systems and methods have taken shape, but there are still some problems. First of all, most album management systems are based on public cloud storage technology. Although the public cloud storage service provider can provide users with convenient massive data storage services, there are huge problems in data security and personal privacy. In recent years, large-scale data leakage incidents occur frequently. Once the sensitive data stored in the cloud platform is leaked, it will bring potential harm to the user's privacy and security. At present, there are a few image management methods based on private cloud storage, especially for personal users or small and micro enterprises. However, in order to control the cost, power consumption, noise, and other factors, the low-noise, low-power, large capacity storage computing resources are generally used, resulting in the lack of computing power, making its album management system only provide simple intelligent image processing functions. In addition, most of the existing album management systems either mainly support the webpage version or mainly support the mobile-end, which leads to high degree of coupling between the functions of front-end and back-end. Programmers cannot easily restructure applications to meet the multi-terminal use needs of users in various application scenarios.

In view of the shortcomings of the existing intelligent album system, this chapter proposes and implements a multi-terminal intelligent album management system and method based on edge computing and private cloud.

1. To protect user data and personal privacy, we build private cloud storage to store images by NAS technology.
2. In order to improve the intelligent level and aesthetic ability of album management system, we integrate edge computing to support intelligent algorithms with high computational power consumption.
3. Based on microservices, we decouple the functions of front-end and back-end, and provide users with convenient multi-end access services for different scenarios.

2 Related Work

2.1 Private Cloud

There are three types of cloud computing: public cloud, private cloud, and hybrid cloud [6]. Hybrid cloud is a combination of public cloud and private cloud. Multiple users of the public cloud share the same cloud server, while the private cloud is built for a single company or individual. Compared with the public cloud, the private cloud can ensure the data security better, because the private cloud is generally deployed in the firewall of the enterprise data center, or in a secure host hosting site. In addition, the private cloud can provide stable service quality without the influence of network instability.

2.2 *Edge Computing*

Edge computing [7] refers to processing data at the edge of the network, that is, the side near the object or data source. Applications launched at the edge of the network can generate faster network service response to meet the basic needs of various industries in real-time business, application intelligence, security, and privacy protection.

In the field of image management, a very important task is to find the object or face in the image. In 2001, the Viola & Jones object detector designed by Paul Viola and Michael Jones achieved higher target detection accuracy and shorter detection time. The Viola & Jones object detector adopts the integral graph which can calculate Haar like features quickly, and adopts the cascade structure and AdaBoost classification method, which greatly improves the detection efficiency [4, 9, 10]. In recent years, with the help of deep learning technology, the field of object detection has been greatly developed. In 2014, Facebook proposed the DeepFace [4], which pioneered the application of the deep convolution neural network in the field of face recognition. The method achieved 97.35% of the face verification accuracy on LFW face dataset, approaching the level of human beings.

Recently, by using some large-scale aesthetic datasets, like AVA [5] and AADB [3], many deep learning methods have been proposed for image-level aesthetic score and aesthetic evaluation [1, 2]. In 2019, Tu et al. [8] proposed an interpretable aesthetic image cropping model, which can find the subgraph with the highest aesthetic evaluation in a picture.

3 Algorithm

Our program consists of three parts: private cloud storage, edge computing, and microservices. They are detailed in Sects. 3.1–3.3. Our overall algorithm flow is shown in Fig. 1.

Step 1: Wait for the user to upload a new picture or submit an intelligent processing instruction. If uploading a new picture, turn to step 2. If it is an intelligent processing instruction, add the picture to the intelligent processing service queue according to the instruction, then go to step 4.

Step 2: Extracts image information, such as image hash value and EXIF information.

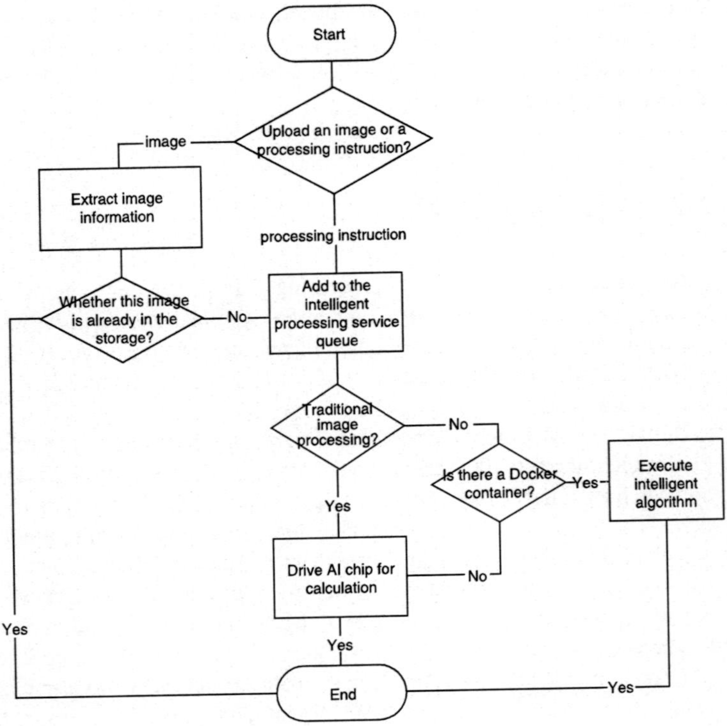

Fig. 1 Algorithm flow chart

Step 3: Judge whether the picture has been put into storage. If not, join the intelligent processing service queue and wait for intelligent analysis. If the picture is already in the storage, the process ends.

Step 4: Check the service queue by the intelligent processing microservice module to see if there are any tasks to be processed. If there are no tasks, keep waiting. If there are tasks, submit them to the corresponding hardware module for processing through the intelligent processing scheduling algorithm.

Step 5: The scheduling algorithm checks the types of image processing tasks. If it is traditional image processing, turns to step 7, otherwise turns to step 6.

Step 6: Check whether there is a registered and running docker container in the system. If so, assign a task and go to step 8. If not, go to step 7.

Step 7: According to the assigned algorithm task, the AI chip is driven to perform the operation. After the operation is completed, the result is returned and stored in the picture information storage.

Step 8: The docker container receives the assigned algorithm task, returns the result after the operation, and stores it in the picture information storage.

3.1 Private Cloud Storage

In our scheme, we use NAS-based private cloud storage system to store images, which is beneficial to protecting users' data and privacy information. The embedded system based on ARM architecture has the advantages of low cost, low power consumption, low noise. and small heat dissipation. In addition, the storage array based on RAID technology ensures the reliability of data.

First, the embedded system based on ARM architecture is selected as the hardware environment of NAS. Then, based on the open source Linux system, we adapt to all the hardware environment, and build a light-weight NAS system suitable for storage, edge computing. and supporting micro service architecture. Finally, we build the storage array module supporting RAID technology, and connect it to the embedded system to provide storage space for private cloud storage.

3.2 Edge Computing

In order to improve the intelligent level of album management method, edge computing is integrated into this scheme. We provide computing services on the data source side to improve response speed and protect user privacy.

We select the appropriate general AI chip and integrate it into the embedded hardware system to provide chip support for edge intelligent computing. In addition, we encapsulate GPU-based intelligent computing framework and its system by Docker container. When the user adds a computer with GPU to the album management system, the system will migrate the encapsulated Docker image to the computer, and then the Docker container can complete the arranged intelligent tasks with the help of the computing power of GPU. After the hardware environment is built, we provide edge intelligence computing in the form of tasks. Before image processing or intelligent computing, the dynamic scheduling algorithm of the system will arrange corresponding hardware resources for these intelligent tasks according to the current resources and load conditions.

3.3 Microservice

Finally, this solution uses the microservices architecture, clearly defines the boundaries of each service module, decouples the front-end and back-end functions, and provides users with multi-terminal access services for different scenarios and a more convenient method to use.

We divide the functions of the system into six modules: user management module, image information management module, intelligent processing module, data storage and management module, system management module, and UI module. We build microservices with these six modules and provide interfaces.

User management module provides user authentication, access authorization, access log, and role management microservices. Only authorized users can access the album resources. At the same time, we adopt role-based access control technology (RBAC), and users can access the storage and computing resources of the system according to their roles.

Image information management module provides microservices related to image information, including the globally unique hash value, EXIF information, and the label attributes defined by users and the information given by intelligent processing module, such as intelligent classification and image quality.

Intelligent processing module provides microservices of intelligent processing. With the help of the computing resources provided by edge intelligent computing, the module performs corresponding intelligent processing for images, such as object detection, face detection, intelligent classification, aesthetic scoring, aesthetic evaluation, and aesthetic image cropping.

Data storage and management module provides microservices for multimedia data storage. Through this module, we can get the multimedia information in the system and manage the storage expansion of the system.

UI module provides unified UI microservices. In order to enable users to have a consistent experience, a unified user interface provided by UI module is shown to users at different terminals, and to get a consistent experience.

4 Experimental Result

The average time taken to perform various intelligent processing on a photo is shown in Table 1. Because the local photo album is tested on a personal laptop with poor hardware configuration, it takes a long time, while the server side with better hardware equipment can process the picture in a shorter time.

Table 1 The average operation time of intelligent processing for the same photo

	Intelligent classification	Aesthetic evaluation	Aesthetic evaluation	Aesthetic image cropping
Webpage end	40 s	20 s	20 s	20 s
Local end	10 h+	5 h	5 h	4 h

5 Conclusion

In conclusion, this chapter proposes a multi-terminal intelligent album management method based on edge computing and private cloud. We make use of private cloud storage to enhance security of users' data and personal privacy. In addition, we integrate edge computing to support intelligent algorithms with high demand for computing power, so as to improve the intelligent level and aesthetic ability of the album management. Finally, we adopt the microservice architecture to meet the needs of users for multiple terminals and various application scenarios.

Acknowledgments This work is partially supported by the National Natural Science Foundation of China (grant numbers 62072014), the Beijing Natural Science Foundation (grant number L192040), the Open Project Program of State Key Laboratory of Virtual Reality Technology and Systems, Beihang University (grant number VRLAB2019C03).

References

1. Dong, Z., Tian, X.: Multi-level photo quality assessment with multi-view features. Neurocomputing **168**, 308–319 (2015)
2. Jin, X., Wu, L., Li, X., Chen, S., Peng, S., Chi, J., Ge, S., Song, C., Zhao, G.: Predicting aesthetic score distribution through cumulative Jensen-Shannon divergence. In: Thirty-Second AAAI Conference on Artificial Intelligence (2018)
3. Kong, S., Shen, X., Lin, Z., Mech, R., Fowlkes, C.: Photo aesthetics ranking network with attributes and content adaptation. In: European Conference on Computer Vision, pp. 662–679. Springer, Cham (2016)
4. Laptev, I.: Improvements of object detection using boosted histograms. In: BMVC, vol. 3, pp. 949–958. Citeseer (2006)
5. Murray, N., Marchesotti, L., Perronnin, F.: Ava: A large-scale database for aesthetic visual analysis. In: 2012 IEEE Conference on Computer Vision and Pattern Recognition, pp. 2408–2415. IEEE, Piscataway (2012)
6. Purcell, B.M.: Big data using cloud computing. J. Technol. Res. **5**, 1 (2014)
7. Satyanarayanan, M.: The emergence of edge computing. Computer **50**(1), 30–39 (2017)
8. Tu, Y., Niu, L., Zhao, W., Cheng, D., Zhang, L.: Image cropping with composition and saliency aware aesthetic score map (2019). Preprint. arXiv: 1911.10492
9. Viola, P., Jones, M.: Rapid object detection using a boosted cascade of simple features. In: Proceedings of the 2001 IEEE computer society conference on computer vision and pattern recognition, CVPR 2001, vol. 1, p. I. IEEE, Piscataway (2001)
10. Viola, P., Jones, M.J.: Robust real-time face detection. Int. J. Comput. Vis. **57**(2), 137–154 (2004)

Author Index

S. Mu et al. (eds.), *4th EAI International Conference on Robotic Sensor Networks*, EAI/Springer Innovations in Communication and Computing,
https://doi.org/10.1007/978-3-030-70451-3

Subject Index

S. Mu et al. (eds.), *4th EAI International Conference on Robotic Sensor Networks*, EAI/Springer Innovations in Communication and Computing,
https://doi.org/10.1007/978-3-030-70451-3

T

U

V

W